ANIMAL TRACKS

of

MINNESOTA & WISCONSIN

Ian Sheldon & Tamara Eder

© 2000 by Lone Pine Publishing
First printed in 2000   10 9 8 7 6 5
Printed in Canada

All rights reserved. No part of this work covered by the copyright hereon may be reproduced or used in any form or by any means—graphic, electronic or mechanical—without the prior written permission of the publisher, except for reviewers, who may quote brief passages. Any request for photocopying, recording, taping or storage in an information retrieval system of any part of this book shall be directed in writing to the publisher.

THE PUBLISHER: LONE PINE PUBLISHING

1808 B Street NW Suite 140	10145-81 Avenue
Auburn, WA 98001	Edmonton, AB T6E 1W9
USA	Canada

Lone Pine Publishing Website: http://www.lonepinepublishing.com

Canadian Cataloguing in Publication Data

Sheldon, Ian, (date)
  Animal tracks of Minnesota and Wisconsin

  Includes bibliographical references and index.
  ISBN 1-55105-250-4

  1. Animal tracks—Minnesota—Identification. 2. Animal tracks—Wisconsin—Identification. I. Eder, Tamara, (date) II. Title.
QL768.S5237 2000   591.47'9   C99-911233-3

Editorial Director: Nancy Foulds
Editor: Volker Bodegom
Proofreader: Randy Williams
Production Manager: Jody Reekie
Design, layout and production: Volker Bodegom, Monica Triska
Cover Design: Rob Weidemann
Cartography: Volker Bodegom
Animal illustrations: Gary Ross, Horst Krause, Ian Sheldon
Track illustrations: Ian Sheldon
Cover illustration: Common Raccoon by Gary Ross
Scanning: Elite Lithographers Ltd.

  We acknowledge the financial support of the Government of Canada through the Book Publishing Industry Development Program (BPIDP) for our publishing activities.

PC: 6

CONTENTS

Introduction **4**
  How to Use This Book **5**
  Map **7**
  Tips on Tracking **8**
  Terms & Measurements . . **10**

Mammals **17**
  Moose 18
  Mule Deer 20
  White-tailed Deer 22
  Horse 24
  Black Bear 26
  Gray Wolf 28
  Coyote 30
  Red Fox 32
  Gray Fox 34
  Lynx 36
  Bobcat 38
  Domestic Cat 40
  Raccoon 42
  Opossum 44
  River Otter 46
  Wolverine 48
  Fisher 50
  Marten 52
  Mink 54
  Long-tailed Weasel 56
  Short-tailed Weasel 58
  Least Weasel 58
  Badger 60
  Striped Skunk 62
  Eastern Spotted Skunk 64
  Snowshoe Hare 66
  Eastern Cottontail 68
  European Hare 70
  Porcupine 72
  Beaver 74
  Muskrat 76
  Woodchuck 78
  Thirteen-lined
    Ground Squirrel 80
  Eastern Chipmunk 82
  Eastern Gray Squirrel 84
  Fox Squirrel 86
  Red Squirrel 88
  Northern Flying Squirrel . . .90
  Norway Rat 92
  Plains Pocket Gopher 94
  Meadow Vole 96
  Deer Mouse 98
  Meadow Jumping Mouse 100
  Masked Shrew 102
  Star-nosed Mole 104

**Birds, Amphibians
  & Reptiles** **107**
  Mallard 108
  Herring Gull 110
  Great Blue Heron 112
  Common Snipe 114
  Spotted Sandpiper 116
  Ruffed Grouse 118
  Great Horned Owl 120
  American Crow 122
  Dark-eyed Junco 124
  Frogs 126
  Toads 128
  Salamanders & Newts 130
  Lizards 132
  Turtles 134
  Snakes 136

Appendix
  Track Patterns & Prints . . **138**
  Print Comparisons **150**
Bibliography **152**
Index **153**

INTRODUCTION

If you have ever spent time with an experienced tracker, or perhaps a veteran hunter, then you know just how much there is to learn about the subject of tracking and just how exciting the challenge of tracking animals can be. Maybe you think that tracking is no fun, because all you get to see is the animal's prints. What about the animal itself—is that not much more exciting? Well, for most of us who don't spend a great deal of time in the beautiful wilderness of Minnesota and Wisconsin, the chances of seeing the majestic Moose or the fun-loving River Otter are slim. The closest that we may ever get to some animals will be through their tracks, and they can inspire a very intimate experience. Remember, you are following in the footsteps of the unseen—animals that are in pursuit of prey, or perhaps being pursued as prey.

Eastern Cottontail

This book offers an introduction to the complex world of tracking animals. Sometimes tracking is easy. At other times it is an incredible challenge that leaves you wondering just what animal made those unusual tracks. Take this book into the field with you, and it can provide some help with the first steps to identification. Animals tracks and trails are this book's focus; you will learn to recognize subtle differences for both. There are, of course, many additional signs to consider, such as scat and food caches, all of which help you to understand the animal that you are tracking.

Remember, it takes many years to become an expert tracker. Tracking is one of those skills that grows with you as you acquire new knowledge in new situations. Most importantly, you will have an intimate experience with nature. You will learn the secrets of the seldom seen. The more you discover, the more you will want to know. And, by developing a good understanding of tracking, you will gain an excellent appreciation of the intricacies and delights of our marvelous natural world.

How to Use This Book

Most importantly, take this book into the field with you! Relying on your memory is not an adequate way to identify tracks. Track identification has to be done in the field or with detailed sketches and notes that you can take home. Much of the process of identification involves circumstantial evidence, so you will have much more success when standing beside the track.

This book is laid out in an easy-to-use format. There is a quick reference appendix to the tracks of all the animals illustrated in the book beginning on p. 138. This appendix is a fast way to familiarize yourself with certain tracks and the content of the book, and it guides you to the more informative descriptions of each animal and its tracks.

Each animal's description is illustrated with the appropriate footprints and the track patterns that it usually leaves. Although these illustrations are not exhaustive, they do show the tracks or groups of prints that you will most likely see. You will find a list of dimensions for the tracks, giving the general range, but there will always be extremes, just as there are with people who have unusually small or large feet. Under the category 'Size' (of animal), the 'greater-than' sign (>) is used when the size difference between the sexes is pronounced.

If you think that you may have identified a track, check the 'Similar Species' section for that animal. This section is designed to help you confirm your conclusions by pointing out other animals that leave similar tracks and showing you ways to distinguish among them.

As you read this book, you will notice an abundance of words such as 'often,' 'mostly' and 'usually.' Unfortunately, tracking will never be an exact science; we cannot expect animals to conform to our expectations, so be prepared for the unpredictable.

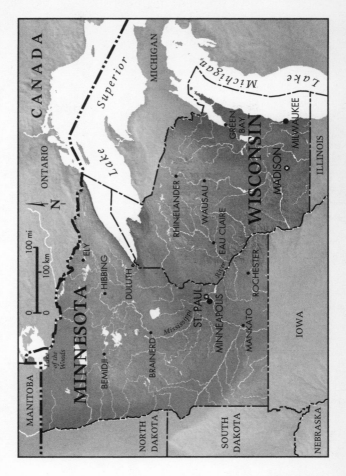

Tips on Tracking

Short-tailed Weasel

As you flip through this guide, you will notice clear, well-formed prints. Do not be deceived! It is a rare track that will ever show so clearly. For a good, clear print, the perfect conditions are slightly wet, shallow snow that isn't melting, or slightly soft mud that isn't actually wet. These conditions can be rare—most often you will be dealing with incomplete or faint prints, where you cannot even be sure of the number of toes.

Should you find yourself looking at a clear print, then the job of identification is much easier. There are a number of key features to look for: Measure the length and width of the print, count the number of toes, check for claw marks and note how far away they are from the body of the print, and look for a heel. Keep in mind more subtle features, such as the spacing between the toes, whether or not they are parallel, and whether fur on the sole of the foot has made the print less clear.

When you are faced with the challenge of identifying an unclear print—or even if you think that you have made a successful identification from one print alone—look beyond the single footprint and search out others. Do not rely on the dimensions of one print alone, but collect measurements from several prints to get an average impression. Even the prints within one trail can show a lot of variation.

Try to determine which is the fore print and which is the hind, and remember that many animals are built very differently from humans, having larger forefeet than hind feet. Sometimes the prints will overlap, or they can be directly on top of one another in a direct register. For some animals, the fore and hind prints are pretty much the same.

Check out the pattern that the tracks make together in the trail, and follow the trail for as many paces as is necessary for you to become familiar with the pattern. Patterns are very important, and they can be the distinguishing feature between different animals with otherwise similar tracks.

Follow the trail for some distance, because it can give you some vital clues. For example, the trail may lead you to a tree, indicating that the animal is a climber—or it may lead down into a burrow. This part of tracking can be the most rewarding, because you are following the life of the animal as it hunts, runs, walks, jumps, feeds or tries to escape a predator.

Take into consideration the habitat. Sometimes very similar species can be distinguished by their habitats only—one might be found on the riverbank, whereas another might be encountered just in the dense forest.

Think about your geographical location, too, because some animals have a limited range. This consideration can rule out some species and help you with your identification.

Remember that every animal will at some point leave a print or trail that looks just like the print or trail of a completely different animal!

Finally, keep in mind that if you track quietly, you might catch up with the maker of the prints.

Terms & Measurements

Some of the terms used in tracking can be rather confusing, and they often depend on personal interpretation. For example, what comes to your mind if you see the word 'hopping'? Perhaps you see a person hopping about on one leg—or perhaps you see a rabbit hopping through the countryside. Clearly, one person's perception of motion can be very different from another's. Some useful terms are explained below, to clarify what is meant in this book, and, where appropriate, how the measurements given fit in with each term.

The following terms are sometimes used loosely and interchangeably—for example, a rabbit might be described as a 'hopper' and a squirrel as a 'bounder,' yet both leave the same pattern of prints in the same sequence.

Ambling: Fast, rolling walking.

Bounding: A gait of four-legged animals in which the two hind feet land simultaneously, usually registering in front of the fore prints. Common in rodents and the rabbit family. 'Hopping' or 'jumping' can often be substituted.

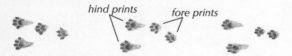

Gait: An animal's gait describes how it is moving at some point in time. Different gaits result in different observable trail characteristics.

Galloping: A gait used by animals with four even-length legs, such as dogs, moving at high speed, hind feet registering in front of forefeet.

Hopping: Similar to bounding. With four-legged animals, usually indicated by tight clusters of prints, fore prints set between and behind the hind prints. A bird hopping on two feet creates a series of paired tracks along its trail.

Loping: Like galloping but slower, with each foot falling independently, and leaving a trail pattern that consists of groups of tracks in the sequence fore-hind-fore-hind, usually roughly in a line.

Mustelids (weasel family) often use ***2×2 loping***, in which the hind feet register directly on the fore prints. The resulting pattern has angled, paired tracks.

Running: Like galloping, but applied generally to animals moving at high speed. Also used for two-legged animals.

Stotting (applies to the Mule Deer only): Describes the action of taking off from the ground and landing on all four feet at once, in pogo-stick fashion.

Trotting: Faster than walking, slower than running. The diagonally opposite limbs move simultaneously; that is, the right forefoot with the left hind, then the left forefoot with the right hind. This gait is the natural one for canids (dog family), short-tailed shrews and voles.

Canids may use ***side-trotting***, a fast trotting in which the hind end of the animal shifts to one side. The resulting track pattern has paired tracks, with all the fore prints on one side and all the hind prints on the other.

Walking: A slow gait in which each foot moves independently of the others, resulting in an alternating track pattern. This gait is common for felines (cat family) and deer, as well as wide-bodied animals, such as bears and porcupines. The term is also used for two-legged animals.

Other Tracking Terms:

Dewclaws: Two small, toe-like structures set above and behind the main foot of most hoofed animals.

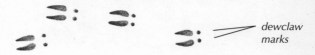

Direct Register: The hind foot falls directly on the fore print.

Double Register: The hind foot registers so as to overlap the fore print only slightly or falls beside it, so that both prints can be seen at least in part.

Dragline: A line left in snow or mud by a foot or the tail dragging over the surface.

Gallop Group: A track pattern of four prints made at a gallop, usually with the hind feet registering in front of the forefeet (see '**galloping**' for illustration).

Height: Taken at the animal's shoulder.

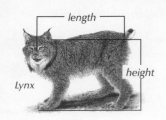

Lynx

Length: The animal's body length from head to rump, not including the tail, unless otherwise indicated.

Metacarpal Pad: A small pad near the palm pad or between the palm pad and heel on the forefeet of bears and members of the weasel family.

Print (also called '*track*'): Fore and hind prints are treated individually. Print dimensions given are 'length' (including claws—maximum values may represent occasional heel register for some animals) and 'width.' A group of prints made by each of the animal's feet makes up a track pattern.

Register: To leave a mark—said about a foot, claw or other part of an animal's body.

Retractable: Describes claws that can be pulled in to keep them sharp, as with the cat family; these claws do not register in the prints. Foxes have semi-retractable claws.

Sitzmark: The mark left on the ground by an animal falling or jumping from a tree.

Straddle: The total width of the trail, all prints considered.

Stride: For consistency among different animals, the stride is taken as the distance from the center of one print (or print group) to the center of the next one. Some books may use the term 'pace.'

Track: Same as '***print***.'

Track Pattern: The pattern left after each foot registers once; a set of prints, such as a gallop group.

Trail: A series of track patterns; think of it as the path of the animal.

Badger

MAMMALS

River Otter

Moose

Fore and Hind Prints
Length: 4–7 in (10–18 cm)
Length with dewclaws: to 11 in (28 cm)
Width: 3.5–6 in (9–15 cm)
Straddle
8.5–20 in (22–50 cm)
Stride
Walking: 1.5–3 ft (45–90 cm)
Trotting: to 4 ft (1.2 m)
Size (bull>cow)
Height: 5–6.5 ft (1.5–2 m)
Length: 7–8.5 ft (2.1–2.6 m)
Weight
600–1100 lb (270–500 kg)

walking

MOOSE
Alces alces

The impressive male Moose, the largest of the deer, has a massive rack of antlers. Moose are usually solitary, though you may see a cow with her calf. Despite its placid appearance, a moose may charge humans if approached.

The ungainly shaped Moose moves gracefully, leaving a neat alternating walking pattern. Its long legs allow for easy movement in snow. The hind prints direct register or double register on the fore prints. Dewclaws—which give extra support for the animal's great weight—register in prints more than 1.2 inches (3 cm) deep, but far behind the hoof. In summer, look for tracks in mud beside ponds and other wet areas, where Moose especially like to feed; they are excellent swimmers. In winter, Moose feed in willow flats and coniferous forests, leaving a distinct browseline (highline). Ripped stems and scraped bark, 6 feet (1.8 m) or more above the ground, are additional signs of Moose.

Similar Species: Deer (pp. 20–23) tracks—smaller, with a narrower straddle—may be mistaken for a juvenile Moose's.

Mule Deer

Fore and Hind Prints
Length: 2–3.3 in (5–8.5 cm)
Width: 1.6–2.5 in (4–6.5 cm)

Straddle
5–10 in (13–25 cm)

Stride
Walking: 10–24 in (25–60 cm)
Stotting: 9–19 ft (2.7–5.8 m)

Size (buck>doe)
Height: 3–3.5 ft (90–110 cm)
Length: 4–6.5 ft (1.2–2 m)

Weight
100–450 lb (45–200 kg)

walking

stot group

MULE DEER
Odocoileus hemionus

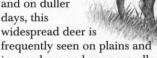

Active primarily in early morning and evening, on moonlit nights and on duller days, this widespread deer is frequently seen on plains and in meadows and open woodlands. In winter it moves down from higher terrain to warmer south-facing slopes and sagebrush flats, where it can still feed without having to contend with deep snow. It prefers small groups and frequently uses the same well-worn path in winter.

The Mule Deer has a neat alternating walking track pattern with the hind prints registered on the fore prints. The prints are heart-shaped and sharply pointed. In deep mud or when the animal is moving quickly, the dewclaws register, closer to the hoof on fore prints than on hind prints. At high speed this deer has a unique gait—stotting—in which it jumps with all four feet leaving and striking the ground at once. Stotting track patterns show how the toes spread to distribute the weight and give better footing.

Similar Species: The White-tailed Deer (p. 22) prefers denser cover and has a different gallop pattern with a shorter gallop stride. Juvenile Moose (p. 18) tracks may be confused with large deer tracks.

White-tailed Deer

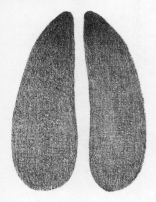

Fore and Hind Prints
Length: 2–3.5 in (5–9 cm)
Width: 1.6–2.5 in (4–6.5 cm)

Straddle
5–10 in (13–25 cm)

Stride
Walking: 10–20 in (25–50 cm)
Galloping: 6–15 ft (1.8–4.5 m)

Size (buck>doe)
Height: 3–3.5 ft (90–110 cm)
Length: to 6.3 ft (1.9 m)

Weight
120–350 lb (55–160 kg)

walking *gallop group*

WHITE-TAILED DEER
Odocoileus virginianus

The keen hearing of this deer guarantees that it knows about you before you know about it. Frequently, all that we see is its conspicuous white tail in the distance as it gallops away, earning this deer the nickname 'flagtail.' This adaptable deer may be found throughout these two states, in small groups at the edges of forests and in brushlands. The White-tailed Deer can be common around ranches and residential areas.

This deer's prints are heart-shaped and pointed. Its alternating walking track pattern shows the hind prints direct registered or double registered on the fore prints. In snow, or when a deer gallops on soft surfaces, the dewclaws register. This flighty deer gallops in the usual style, leaving hind prints ahead of fore prints, with toes spread wide for better footing.

Similar Species: Mule Deer (p. 20) prints are almost identical, but they stot (not gallop) and have a different habitat. Juvenile Moose (p. 18) tracks may be confused with large deer tracks.

Horse

**Fore Print
(hind print is slightly smaller)**
Length: 4.5–6 in (11–15 cm)
Width: 4.5–5.5 in (11–14 cm)

Straddle
2–7.5 in (5–19 cm)

Stride
Walking: 17–28 in (43–70 cm)

Size
Height: to 6 ft (1.8 m)

Weight
to 1500 lb (680 kg)

walking

HORSE
Equus caballus

Back-country use of the popular Horse means that you can expect its tracks to show up almost anywhere.

Unlike any other animal in this book, the Horse has only one huge toe. This toe leaves an oval print with a distinctive 'frog' (V-shaped mark) at its base. When the Horse is shod, the horseshoe shows up clearly as a firm wall at the outside of the print. Not all horses are shod, so do not expect to see this outer wall on every horse print. A typical, unhurried horse trail shows an alternating walking pattern, with the hind prints registered on or behind the slightly larger fore prints. Horses are capable of a range of speeds—up to a full gallop—but most recreational horseback riders take a more leisurely outlook on life, preferring to walk their horses and soak up the views!

Similar Species: Mules (rarely shod) have smaller tracks.

Black Bear

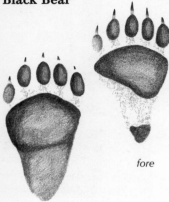

fore

hind

Fore Print
Length: 4–6.3 in (10–16 cm)
Width: 3.8–5.5 in (9.5–14 cm)
Hind Print
Length: 6–7 in (15–18 cm)
Width: 3.5–5.5 in (9–14 cm)
Straddle
9–15 in (23–38 cm)
Stride
Walking: 17–23 in (43–58 cm)
Size (male>female)
Height: 3–3.5 ft (90–110 cm)
Length: 5–6 ft (1.5–1.8 m)
Weight
200–600 lb (90–270 kg)

walking (slow)

BLACK BEAR
Ursus americanus

The Black Bear is widespread in forested areas throughout Minnesota and Wisconsin, but do not expect to see its tracks in winter, when it sleeps deeply. Finding fresh bear tracks can be a thrill, but take care—the bear may be just ahead. Never underestimate the potential power of a surprised bear!

Black Bear prints resemble small human prints, but they are wider and show claw marks. The small inner toe rarely registers. The forefoot's small heel pad often shows in the print, and the hind print shows a big heel. The bear's slow walk results in a slightly pigeon-toed double register with the hind print on the fore print. More frequently, at a faster pace, the hind foot oversteps the forefoot. When a bear runs, the two hind feet register in front of the forefeet in an extended cluster. Along well-worn bear paths, look for 'digs'—patches of dug-up earth—and 'bear trees' whose scratched bark shows that these bears climb.

Similar Species: The magnificent Grizzly (Brown) Bear (*Ursus arctos*) has similar prints but no longer lives here.

Gray Wolf

fore

hind

**Fore Print
(hind print is slightly smaller)**
Length: 4–5.5 in (10–14 cm)
Width: 2.5–5 in (6.5–13 cm)

Straddle
3–7 in (7.5–18 cm)

Stride
Walking: 15–32 in (38–80 cm)
Galloping: 3 ft (90 cm)
    leaps to 9 ft (2.7 m)

Size (female is slightly smaller)
Height: 25–37 in (65–95 cm)
Length: 3.5–5.3 ft (1.1–1.6 m)

Weight
70–120 lb (32–55 kg)

walking   *trotting*

GRAY WOLF
(Timber Wolf)
Canis lupus

The soulful howl of the wolf epitomizes the outdoor experience, but few people ever hear it—your best bet is in remote, undisturbed areas. The rarely seen Gray Wolf, the largest of the wild dogs, may travel in packs or alone.

The Gray Wolf leaves a straight alternating walking track pattern of large, oval prints that each show all four claws. The smaller hind feet register directly on the prints made by the larger forefeet. Note the different lobing on the fore and hind heel pads. In deep snow, wolves sensibly follow their leader's trail, sometimes dragging their feet. In a trotting track pattern, the hind print has a slight lead and falls to one side, giving an unbalanced appearance. Wolves and Coyotes (p. 30) gallop in the same manner.

Similar Species: Domestic Dog (*Canis familiaris*) prints, rarely as large as a wolf's, fall in a haphazard track pattern with a less direct register, and the inner toes tend to splay more. Coyote tracks are much smaller. A four-toed print by a Wolverine (p. 48) may be confused with a wolf's, but the pad shapes are very different.

Coyote

fore

hind

**Fore Print
(hind print is slightly smaller)**
Length: 2.4–3.2 in (6–8 cm)
Width: 1.6–2.4 in (4–6 cm)

Straddle
4–7 in (10–18 cm)

Stride
Walking: 8–16 in (20–40 cm)
Trotting: 17–23 in (43–58 cm)
Galloping/Leaping:
   2.5–10 ft (0.8 m–3 m)

Size (female is slightly smaller)
Height: 23–26 in (58–65 cm)
Length: 32–40 in (80–100 cm)

Weight
20–50 lb (9–23 kg)

walking or trotting   *gallop group*

COYOTE (Brush Wolf, Prairie Wolf)
Canis latrans

This widespread, adaptable canine prefers open grasslands or woodlands. It hunts rodents and larger prey, either on its own, with a mate or in a family pack. A Coyote occasionally develops an interesting cooperative relationship with a Badger (p. 60), so you might find their tracks together where they have been digging for ground squirrels (p. 80).

The oval fore prints are slightly larger than the hind prints. The fore heel pad is more triangular than the hind heel pad, which rarely registers clearly. The two outer toes usually do not register claw marks. The Coyote typically walks or trots in an alternating pattern; the walk has a wider straddle, and the trotting trail is often very straight. When it gallops, the Coyote's hind feet fall in front of its forefeet; the faster it goes, the straighter the gallop group. The Coyote's tail, which hangs down, leaves a dragline in deep snow.

Similar Species: A Domestic Dog's (*Canis familiaris*) less-oval prints splay more, and its trail is erratic. Foot hairs make Red Fox (p. 32) prints (usually smaller) less clear. Gray Fox (p. 34) tracks are much smaller.

Red Fox

fore

hind

Fore Print
(hind print is slightly smaller)
Length: 2.1–3 in (5.3–7.5 cm)
Width: 1.6–2.3 in (4–5.8 cm)

Straddle
2–3.5 in (5–9 cm)

Stride
Trotting: 12–18 in (30–45 cm)
Side-trotting: 14–21 in (36–53 cm)

Size (vixen is slightly smaller)
Height: 14 in (35 cm)
Length: 22–25 in (55–65 cm)

Weight
7–15 lb (3.2–7 kg)

trotting *side-trotting*

RED FOX
Vulpes vulpes

This beautiful and notoriously cunning fox, found throughout Minnesota and Wisconsin, prefers mountainous forests and open areas. It is a very adaptable and intelligent animal.

The finer details of a Red Fox's tracks are blurred by its foot hairs, so only parts of the toes and heel pads show. The horizontal or slightly curved bar across the fore heel pad is diagnostic. A trotting Red Fox leaves a distinctive, straight trail of alternating prints—the hind print direct registers on the wider fore print. When a fox side-trots, it leaves print pairs in which the hind print falls to one side of the fore print in typical canid fashion. Foxes gallop like Coyotes (p. 30). The faster the gallop, the straighter the gallop group.

Similar Species: Other canid prints lack the bar across the fore heel pad. Gray Fox (p. 34) prints are smaller. Domestic Dog (*Canis familiaris*) prints can be of similar size. Small Coyote prints are similar, but they have a wider straddle, and the toe marks are more bulbous.

Gray Fox

fore

hind

Fore Print
(hind print slightly smaller)
Length: 1.3–2.1 in (3.3–5.3 cm)
Width: 1.1–1.5 in (2.8–3.8 cm)

Straddle
2–4 in (5–10 cm)

Stride
Walking/Trotting: 10–12 in (25–30 cm)

Size
Height: 14 in (35 cm)
Length: 21–30 in (53–75 cm)

Weight
7–15 lb (3.2–7 kg)

walking

GRAY FOX
Urocyon cinereoargenteus

This small, shy fox is widespread, but it especially prefers woodlands and chaparral country. The Gray Fox is the only fox that climbs trees, which it does either for safety or to forage.

The forefoot registers better than the smaller hind foot, and the hind foot's long, semi-retractable claws do not always register. The heel pads are often unclear—they sometimes show up just as small, round dots. When it walks, this fox leaves a neat alternating track pattern; when it trots, its prints fall in pairs, with the fore print set diagonally behind the hind print. The Gray Fox's gallop group is like the Coyote's (p. 30).

Similar Species: The Red Fox (p. 32) has heel pads with a bar across them; in general, its prints are larger and less clear (because of thick fur), its stride is longer and its straddle is narrower. Coyote tracks are much larger. Domestic Cat (p. 40) and Bobcat (p. 38) prints lack claw marks and have larger, less symmetrical heel pads.

Lynx

fore

hind

Fore Print
(hind print is slightly smaller)
Length: 3.5–4.5 in (9–11 cm)
Width: 3.5–4.8 in (9–12 cm)

Straddle
6–9 in (15–23 cm)

Stride
Walking: 12–28 in (30–70 cm)

Size
Length: 2.5–3 ft (75–90 cm)

Weight
15–30 lb (7–14 kg)

walking

LYNX
Lynx canadensis

This large cat is a thrill to see, but it is sensitive to human interference. The elusive Lynx is abundant only in the remote, undisturbed, dense forests of northern Minnesota and Wisconsin. With its huge feet and a relatively lightweight body, it stays on top of the snow as it pursues its main prey, the Snowshoe Hare (p. 66).

This cautious walker leaves a neat alternating track pattern, the hind print direct registered on top of the fore print. Thick fur on the feet often results in prints that are big, round depressions with no detail. In deeper snow the print may be extended by 'handles' off to the rear. However deep the snow, this cat sinks no more than 8 inches (20 cm), and it rarely drags its feet. The Lynx is more likely to bound than to run. Its curious nature results in a meandering trail that may lead to a partially buried food cache.

Similar Species: Bobcat (p. 38) prints are smaller, and the trail may show draglines. Fisher (p. 50) prints may not show the fifth toe—look for mustelid habits. Coyote (p. 30), Fox (pp. 32–35) and Domestic Dog (*Canis familiaris*) prints show claw marks, their length exceeds their width and the front of the footpad has just one lobe.

Bobcat

fore

hind

Fore Print
(hind print is slightly smaller)
Length: 1.8–2.5 in (4.5–6.5 cm)
Width: 1.8–2.5 in (4.5–6.5 cm)

Straddle
4–7 in (10–18 cm)

Stride
Walking: 8–16 in (20–40 cm)
Running: 4–8 ft (1.2–2.4 m)

Size
(female is slightly smaller)
Height: 20–22 in (50–55 cm)
Length: 25–30 in (65–75 cm)

Weight
15–35 lb (7–16 kg)

walking   *ambling to loping*

BOBCAT (Wildcat)
Lynx rufus

The widely distributed Bobcat, a stealthy and usually nocturnal hunter, is seldom seen. Very adaptable, it can leave tracks anywhere from wild mountainsides to chaparral and even into residential areas.

When a Bobcat walks, its hind feet usually register directly on the larger fore prints. As a Bobcat picks up speed, its trail becomes an ambling pattern of paired prints, the hind leading the fore. At even greater speeds, it leaves four-print groups in a lope pattern. Especially the fore prints show asymmetry. The front part of the heel pad has two lobes and the rear part has three. In deep snow the Bobcat's feet leave draglines. The Bobcat marks its territory with half-buried scat along its meandering trail.

Similar Species: Large Domestic Cats (p. 40) have similar prints with a shorter stride and a narrower straddle. Coyote (p. 30), Fox (pp. 32–35) and Domestic Dog (*Canis familiaris*) tracks are narrower than they are long and show claw marks, and the fronts of their footpads are once-lobed; wild canid trails do not meander. Some mustelids' (pp. 50–56) hind prints may also seem similar—look for mustelid habits.

Domestic Cat

fore

hind

**Fore Print
(hind print is slightly smaller)**
Length: 1–1.6 in (2.5–4 cm)
Width: 1–1.8 in (2.5–4.5 cm)

Straddle
2.4–4.5 in (6–11 cm)

Stride
Walking: 5–8 in (13–20 cm)
Loping/Galloping:
 14–32 in (35–80 cm)

Size (male>female)
Height: 20–22 in (50–55 cm)
Length with tail: 30 in (75 cm)

Weight
6.5–13 lb (3–6 kg)

walking

*loping
to galloping*

DOMESTIC CAT
(House Cat)
Felis catus

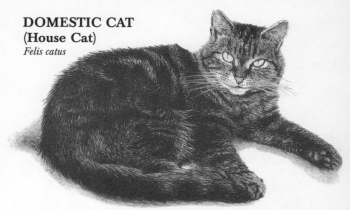

The tracks of the familiar and abundant Domestic Cat can show up almost any place where there are people. Abandoned cats may roam farther afield, and these 'feral cats' lead a pretty wild and independent existence. Domestic Cats can come in many shapes, sizes and colors.

As with all felines, the Domestic Cat's fore print and slightly smaller hind print both show four toe pads. Its retractable claws, kept clean and sharp for catching prey, do not register. Cat prints usually show a slight asymmetry, with one toe leading the others. A Domestic Cat makes a neat alternating walking track pattern, usually in direct register, as one would expect from this animal's fastidious nature. When a cat picks up speed, it leaves clusters of four prints, the hind feet registering in front of the forefeet.

Similar Species: A small Bobcat (p. 38) may leave tracks similar to a very large Domestic Cat's. Fox (pp. 32–35) and Domestic Dog (*Canis familiaris*) prints show claw marks.

Raccoon

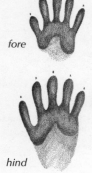

fore

hind

Fore Print
Length: 2–3 in (5–7.5 cm)
Width: 1.8–2.5 in (4.5–6.5 cm)

Hind Print
Length: 2.4–3.8 in (6–9.5 cm)
Width: 2–2.5 in (5–6.5 cm)

Straddle
3.3–6 in (8.5–15 cm)

Stride
Walking: 8–18 in (20–45 cm)
Bounding: 15–25 in (38–65 cm)

Size
(female is slightly smaller)
Length: 24–37 in (60–95 cm)

Weight
11–35 lb (5–16 kg)

walking   *bounding group*

RACCOON
Procyon lotor

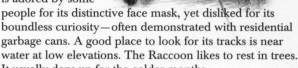

The inquisitive Raccoon, common in this region, is adored by some people for its distinctive face mask, yet disliked for its boundless curiosity—often demonstrated with residential garbage cans. A good place to look for its tracks is near water at low elevations. The Raccoon likes to rest in trees. It usually dens up for the colder months.

The Raccoon's unusual print, showing five well-formed toes, looks like a human handprint; its small claws make dots. Its highly dexterous forefeet rarely leave heel prints, but its hind prints, which are generally much clearer, do show heels. The Raccoon's peculiar walking track pattern shows the left fore print next to the right hind print (or just in front) and vice versa. On the rare occasions when a Raccoon is out in deep snow, it may use a direct-registering walk. The Raccoon occasionally bounds, leaving clusters with the hind prints in front of the fore prints.

Similar Species: Unclear Opossum (p. 44) prints may look similar, but the Opossum drags its tail. In deep snow, Fisher (p. 50) or River Otter (p. 46) tracks may look similar.

Opossum

fore

hind

Fore Print
Length: 2–2.3 in (5–5.8 cm)
Width: 2–2.3 in (5–5.8 cm)
Hind Print
Length: 2.5–3 in (6.5–7.5 cm)
Width: 2–3 in (5–7.5 cm)
Straddle
4–5 in (10–13 cm)
Stride
Walking: 5–11 in (13–28 cm)
Size
Length: 2–2.5 ft (60–75 cm)
Weight
9–13 lb (4–6 kg)

walking *walking (fast)*

OPOSSUM
Didelphis virginiana

This slow-moving, nocturnal marsupial is found in all but the extreme northern parts of Minnesota and Wisconsin. Though it occupies many habitats, it prefers open woodland or brushland around waterbodies. It is quite tolerant of residential areas. The Opossum's tracks can often be seen in mud near the water and in snow during the warmer months of winter, but it tends to den up during severe weather.

The Opossum is an excellent climber, so its trail may lead to a tree. It has two walking habits: the common alternating pattern, with the hind prints registering on the fore prints, and a Raccoon-like (p. 42) paired-print pattern, with each hind print next to the opposing fore print. The very distinctive, long, inward-pointing thumb of the hind foot does not make a claw mark. In snow, the dragline of the long, naked tail may be stained with blood; unfortunately, this thinly haired animal is not well adapted to cold and frequently suffers frostbite.

Similar Species: Prints in which the distinctive thumbs do not show, as in sand or fine snow, may be mistaken for a Raccoon's.

River Otter

fore

hind

Fore Print
Length: 2.5–3.5 in (6.5–9 cm)
Width: 2–3 in (5–7.5 cm)

Hind Print
Length: 3–4 in (7.5–10 cm)
Width: 2.3–3.3 in (5.8–8.5 cm)

Straddle
4–9 in (10–23 cm)

Stride
Loping: 12–27 in (30–70 cm)

**Size
(female is two-thirds the size of male)**
Length with tail: 3–4.3 ft (90–130 cm)

Weight
10–25 lb (4.5–11 kg)

loping (fast)

RIVER OTTER
Lontra canadensis

No animal knows how to have more fun than a River Otter. If you are lucky enough to watch one at play, you will not soon forget the experience. Widespread and well-adapted for the aquatic environment, this otter lives along waterbodies; an otter in the forest is usually on its way to another waterbody. Expect to see a wealth of evidence of an otter's presence along the riverbanks in its home territory.

In soft mud, the River Otter's five-toed feet, especially the hind ones, register evidence of webbing. The inner toes are set slightly apart. If the forefoot's metacarpal pad registers, it lengthens the print. Very variable, otter trails usually show a typical mustelid 2×2 loping. However, with faster gaits they show groups of four and three prints. The thick, heavy tail often leaves a dragline. This otter loves to slide in snow, often down riverbanks, leaving troughs nearly 1 foot (30 cm) wide. In summer it rolls and slides on the grass and in the mud.

Similar Species: Other mustelid trails lack conspicuous tail drag. The Marten (p. 52) has similar-sized prints. The Fisher (p. 50) usually has hairy feet, with the forefeet larger than the hind feet. Mink (p. 54) prints are about half the size.

Wolverine

fore

hind

Fore Print
Length with heel: 4–7.5 in (10–19 cm)
Width: 4–5 in (10–13 cm)

Hind Print
Length: 3.5–4 in (9–10 cm)
Width: 4–5 in (10–13 cm)

Straddle
7–9 in (18–23 cm)

Stride
Walking: 3–12 in (7.5–30 cm)
Running: 10–40 in (25–100 cm)

Size (female is slightly smaller)
Height: 16 in (40 cm)
Length: 2.7–3.8 ft (80–120 cm)

Weight
18–45 lb (8–21 kg)

loping (slow)

WOLVERINE
Gulo gulo

The reputation of the robust and powerful Wolverine, the largest of the mustelids, has earned it many nicknames, such as 'skunk bear' and 'Indian devil.' The Wolverine lives in coniferous or mixed forests. Its need for pristine wilderness has resulted in a scattered distribution throughout much of its range.

Like other mustelids, the Wolverine has five-toed feet, but the inner toe rarely registers in the print. Though the forefoot registers a small heel pad, the hind foot rarely does. Because of its low, squat shape, the Wolverine leaves a host of erratic typical mustelid trails: an alternating walking pattern, the typical 2×2 loping pattern with its print pairs, and the common loping pattern of three- and four-print groups.

Similar Species: Most other mustelids (pp. 46–61) make much smaller prints, and their track patterns are less erratic. When only four toes register, Wolverine tracks may be mistaken for Gray Wolf (p. 28) or Domestic Dog (*Canis familiaris*) tracks, but the pad shapes are quite different.

Fisher

fore

Fore Print
Length: 2.1–4 in (5.3–10 cm)
Width: 2.1–3.3 in (5.3–8.5 cm)

Hind Print
Length: 2.1–3 in (5.3–7.5 cm)
Width: 2–3 in (5–7.5 cm)

Straddle
3–7 in (7.5–18 cm)

Stride
Walking: 7–14 in (18–35 cm)
2×2 loping: 1–4.3 ft (30–130 cm)
Loping: 1–3 ft (30–90 cm)

Size (male>female)
Length with tail:
   34–40 in (85–100 cm)

Weight
3–12 lb (1.4–5.5 kg)

walking   *2×2 loping*

FISHER (Black Cat)
Martes pennanti

This agile hunter is comfortable both on the ground and in the trees of mixed hardwood forests. The Fisher's speed and eager hunting antics make for exciting tracking as it races up trees and along the ground in its quest for squirrels. It is one of the few predators to kill and eat Porcupines (p. 72).

Though all five toes may register, the small inner toe frequently does not. Only the forefoot has a small heel pad that can show up in the print. The Fisher occasionally walks, making a direct-registering alternating track pattern, but it more often 2×2 lopes in typical mustelid fashion, leaving angled print pairs with the hind print direct registered on the fore print. Loping, its most common gait, produces three- and four-print groups (see the River Otter, p. 46). The patterns often vary within a short distance. Fishers are not associated with water, so 'Fisher' is a misnomer!

Similar Species: Male Marten (p. 52) tracks may be confused with a small female Fisher's, but Martens weigh less and leave shallower prints. Otters have larger hind feet than forefeet. When only four toes register, Fisher prints may look like a Bobcat's (p. 38).

Marten

Fore and Hind Prints
Length: 1.8–2.5 in (4.5–6.5 cm)
Width: 1.5–2.8 in (3.8–7 cm)

Straddle
2.5–4 in (6.5–10 cm)

Stride
Walking: 4–9 in (10–23 cm)
2×2 loping: 9–46 in (23–120 cm)

Size (male>female)
Length with tail:
    21–28 in (53–70 cm)

Weight
1.5–2.8 lb (0.7–1.3 kg)

walking   *2×2 loping*

MARTEN
(American Sable)
Martes americana

This aggressive predator is found in the coniferous and mixed-wood forests of Minnesota and Wisconsin.

The Marten seldom leaves a clear print: often just four toes register and the heel pad is undeveloped. In winter the hairiness of the feet often blurs all pad detail, especially the poorly developed palm pads. In the Marten's alternating walk, the hind feet register on the fore prints. In 2×2 loping, the hind prints fall on the fore prints to form slightly angled print pairs in a typical mustelid pattern. Its loping track patterns may appear as three- or four-print clusters (see the River Otter, p. 46). Follow the criss-crossing trails—if a Marten has scrambled up a tree, look for a sitzmark where it has jumped down.

Similar Species: Size and habitat are often key to distinguishing Marten, Fisher (p. 50) and Mink (p. 54) tracks. A female Fisher's prints may resemble a large male Marten's, but they will be clearer. Male Mink prints overlap in size with small female Marten prints, but Mink rarely climb trees and (unlike Martens) are usually found near water.

Mink

fore

hind

Fore and Hind Prints
Length: 1.3–2 in (3.3–5 cm)
Width: 1.3–1.8 in (3.3–4.5 cm)

Straddle
2.1–3.5 in (5.3–9 cm)

Stride
Walking/2×2 loping: 8–36 in (20–90 cm)

Size (male>female)
Length with tail: 19–28 in (48–70 cm)

Weight
1.5–3.5 lb (0.7–1.6 kg)

2x2 loping

MINK
Mustela vison

The lustrous Mink, widespread throughout the region, prefers watery habitats surrounded by brush or forest. At home as much on land as in water, this nocturnal hunter can be exciting to track. Like the River Otter (p. 46), the Mink sometimes slides in snow, carving out a trough up to 6 inches (15 cm) wide for an observant tracker to spot.

The Mink's fore print shows five (perhaps four) toes and five loosely connected palm pads in an arc, but the hind print shows only four palm pads. The metacarpal pad of the forefoot rarely registers, but the furred heel of the hind foot may register, lengthening the hind print. The Mink prefers the typical mustelid 2×2 loping, making consistently spaced, slightly angled double prints. Its diverse track patterns also include alternating walking, loping with three- and four-print groups (like the River Otter) and bounding (like rabbits and hares, pp. 66–71).

Similar Species: Small Martens (p. 52) may have similar prints, but without a consistent 2×2 loping gait, and they do not live near water. Weasels (pp. 56–59) make similar but smaller tracks.

Weasels

all weasels

Long-tailed Weasel
Fore and Hind Prints
Length: 1.1–1.8 in (2.8–4.5 cm)
Width: 0.8–1 in (2–2.5 cm)

Straddle
1.8–2.8 in (4.5–7 cm)

Stride
2×2 loping: 9.5–43 in (24–110 cm)

Size (male>female)
Length with tail: 12–22 in (30–55 cm)

Weight
3–12 oz (85–340 g)

*2×2 loping
(Long-tailed)*

LONG-TAILED WEASEL
Mustela frenata

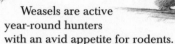

Weasels are active year-round hunters with an avid appetite for rodents. The Long-tailed Weasel, the largest of the three weasels in Minnesota and Wisconsin, is widely distributed.

Following a weasel's tracks can reveal much about the activities of the nimble creature. Tracks are most evident in winter, when weasels frequently burrow into the snow or pursue rodents into their holes. Some weasel trails may lead you up a tree. Weasels sometimes take to water. To identify the weasel species, pay close attention to the straddle, stride and loping patterns, and note the distribution and habitat. The usual weasel gait is a 2×2 lope, leaving a trail of paired prints. A weasel's light weight and small, hairy feet result in pad detail that is often unclear, especially in snow. Even with clear tracks, the inner (fifth) toe rarely registers.

The Long-tailed Weasel's typical 2×2 lope shows an irregular stride—sometimes short and sometimes long—with no consistent behavior. Like the Mink (p. 54), this weasel may bound like a rabbit or hare (pp. 66–71).

Similar Species: A large male Short-tailed Weasel's (p. 58) tracks may be the same size as a small female Long-tailed Weasel's. The Least Weasel (p. 58) shares some habitats with the Long-tailed Weasel, but it has much smaller tracks.

Weasels

Short-tailed Weasel
Fore and Hind Prints
Length: 0.8–1.3 in (2–3.3 cm)
Width: 0.5–0.6 in (1.3–1.5 cm)

Straddle
1–2.1 in (2.5–5.3 cm)

Stride
2×2 loping: 9–36 in (23–90 cm)

Size (male>female)
Length with tail:
    8–14 in (20–35 cm)

Weight
1–6 oz (30–170 g)

Least Weasel
Fore and Hind Prints
Length: 0.5–0.8 in (1.3–2 cm)
Width: 0.4–0.5 in (1–1.3 cm)

Straddle
0.8–1.5 in (2–3.8 cm)

Stride
2×2 loping: 5–20 in (13–50 cm)

Size (male>female)
Length with tail:
    6.5–9 in (17–23 cm)

Weight
1.3–2.3 oz (37–65 g)

2×2 loping (Short-tailed)   *2×2 loping (Least)*

SHORT-TAILED WEASEL
(Ermine, Stoat)
Mustela erminea

The widely distributed Short-tailed Weasel prefers woodlands and meadows up to higher elevations, but it does not favor wetlands or dense coniferous forests. This weasel's 2×2 loping tracks may fall in clusters, with alternating short and long strides.

Similar Species: Small female Long-tailed Weasel (p. 56) tracks may be the same size as a large male Short-tailed's. Large male Least Weasel (below) tracks may be the same size as a small female Short-tailed's.

LEAST WEASEL
Mustela nivalis

The Least Weasel, found throughout Minnesota and Wisconsin, is the smallest weasel, with the least-clear tracks. Its tracks may be found around wetlands and in open woodlands and fields.

Similar Species: A small female Short-tailed Weasel's (above) tracks may resemble a large male Least Weasel's, but Short-tailed Weasels do not frequent wet areas, preferring upland areas and woodlands.

Badger

fore

hind

**Fore Print
(hind print is slightly smaller)**
Length: 2.5–3 in (6.5–7.5 cm)
Width: 2.3–2.8 in (5.8–7 cm)

Straddle
4–7 in (10–18 cm)

Stride
Walking: 6–12 in (15–30 cm)

Size
Length: 21–36 in (53–90 cm)

Weight
13–25 lb (6–11 kg)

walking

BADGER
Taxidea taxus

The squat shape and unmistakable face of this bold animal are most often seen in the open grasslands, but the Badger also ventures into high mountain country. They are found throughout most of Minnesota and Wisconsin. Thick shoulders and forelegs, coupled with long claws, make the Badger a powerful digger. Look for tracks in spring and fall snow—unlike most other mustelids, the Badger likes to den up in a hole for the coldest months of winter.

When a Badger walks, the alternating track pattern shows a double register, with the hind print sometimes falling just behind (or sometimes slightly in front of) the fore print. All five toes on each foot register. A Badger's long claws are evident in the pigeon-toed track that it leaves as it waddles along; the forefoot claws are longer than the hindfoot ones. This animal's wide, low body often plows through deeper snow, obscuring track detail.

Similar Species: In snow, a Porcupine's (p. 72) trail may be similar, but it will show draglines made by the tail and quills, and it will likely lead up a tree, not to a hole.

Striped Skunk

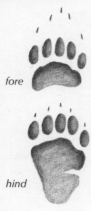

fore

hind

Fore Print
Length: 1.5–2.2 in (3.8–5.6 cm)
Width: 1–1.5 in (2.5–3.8 cm)

Hind Print
Length: 1.5–2.5 in (3.8–6.5 cm)
Width: 1–1.5 in (2.5–3.8 cm)

Straddle
2.8–4.5 in (7–11 cm)

Stride
Walking/Bounding:
 2.5–8 in (6.5–20 cm)

Size
Length with tail:
 20–32 in (50–80 cm)

Weight
6–14 lb (2.7–6.5 kg)

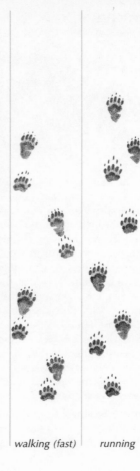

walking (fast)   *running*

STRIPED SKUNK
Mephitis mephitis

This striking skunk has a notorious reputation for its vile smell, and the lingering odor is often the best sign of its presence. Widespread throughout the region in a diversity of habitats, it prefers lower elevations. The Striped Skunk dens up in winter, coming out on warmer days and in spring.

Forefeet and hind feet each have five toes. The long claws on the forefeet often register. The smooth palm pads and small heel pads leave surprisingly small prints. Skunks mostly walk—with such a potent smell for their defense, and those memorable black and white stripes, they rarely need to run. Note that their trail rarely shows any consistent pattern, though an alternating walking pattern may be evident. The greater a skunk's speed, the more the hind foot oversteps the fore. If it runs, its trail consists of clumsy, closely set four-print groups. In snow it drags its feet.

Similar Species: The Eastern Spotted Skunk (p. 64), found in Minnesota and western Wisconsin, makes smaller prints in a very random pattern. Mustelid (pp. 46–61) tracks are farther apart than those resulting from a skunk's shuffling gait, and this skunk's prints do not overlap.

Eastern Spotted Skunk

fore

hind

Fore Print
Length: 1–1.3 in (2.5–3.3 cm)
Width: 0.9–1.1 in (2.3–2.8 cm)

Hind Print
Length: 1.2–1.5 in (3–3.8 cm)
Width: 0.9–1.1 in (2.3–2.8 cm)

Straddle
2–3 in (5–7.5 cm)

Stride
Walking: 1.5–3 in (3.8–7.5 cm)
Bounding: 6–12 in (15–30 cm)

Size
Length: 13–25 in (33–65 cm)

Weight
0.6–2.2 lb (0.3–1 kg)

walking

bounding

EASTERN SPOTTED SKUNK
Spilogale putorius

This beautifully marked skunk, smaller than its striped cousin, is found throughout Minnesota and in western Wisconsin. It enjoys diverse habitats—such as scrubland, forests and farmland—but it is a rare sight, because of its nocturnal habits and because it dens up in winter, coming out only on warmer nights.

This skunk leaves a very haphazard trail as it forages for food on the ground. Occasionally, and with ease, it climbs trees. The long claws on the forefeet often register, and the palm and heel may leave defined pad marks. Although this skunk rarely runs, when it does so it may bound along, leaving groups of four prints, hind in front of fore. It sprays only when truly provoked, so its powerful odor is less frequently detected than that of the Striped Skunk (p. 62).

Similar Species: The Striped Skunk has larger prints and less-scattered tracks with a shorter running stride (or it jumps); it does not climb trees.

Snowshoe Hare

hind   *fore*

Fore Print
Length: 2–3 in (5–7.5 cm)
Width: 1.5–2 in (3.8–5 cm)
Hind Print
Length: 4–6 in (10–15 cm)
Width: 2–3.5 in (5–9 cm)
Straddle
6–8 in (15–20 cm)
Stride
Hopping: 0.8–4.3 ft (25–130 cm)
Size
Length: 12–21 in (30–53 cm)
Weight
2–4 lb (0.9–1.8 kg)

hopping

SNOWSHOE HARE
(Varying Hare)
Lepus americanus

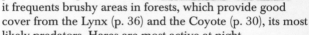

This hare is well known for its color change from summer brown to winter white and for its huge hind feet, which enable it to 'float' on top of snow. Widespread throughout the northern parts of this region, it frequents brushy areas in forests, which provide good cover from the Lynx (p. 36) and the Coyote (p. 30), its most likely predators. Hares are most active at night.

As with other hares and rabbits, the Snowshoe Hare's most common track pattern is a hopping one, with triangular four-print groups; they can be quite long if the hare moves quickly. In winter, heavy fur on the hind feet (much larger than the forefeet) thickens the toes, which can splay out to further distribute the hare's weight on snow. Hares make well-worn runways that are often used as escape runs. You may encounter a resting hare, because hares do not live in burrows. Twigs and stems neatly cut at a 45° angle also indicate this hare's presence.

Similar Species: Eastern Cottontail (p. 68) prints are similar, but smaller. The White-tailed Jackrabbit (*Lepus townsendii*) of Minnesota splays its hind toes less.

Eastern Cottontail

hind

fore

Fore Print
Length: 1–1.5 in (2.5–3.8 cm)
Width: 0.8–1.3 in (2–3.3 cm)
Hind Print
Length: 3–3.5 in (7.5–9 cm)
Width: 1–1.5 in (2.5–3.8 cm)
Straddle
4–5 in (10–13 cm)
Stride
Hopping: 0.6–3 ft (18–90 cm)
Size
Length: 12–17 in (30–43 cm)
Weight
1.3–3 lb (0.6–1.4 kg)

hopping

EASTERN COTTONTAIL
Sylvilagus floridanus

This abundant rabbit, widespread throughout Minnesota and Wisconsin, prefers brushy areas in grasslands and cultivated areas. It might be found in dense vegetation, where it hides from predators such as the Bobcat (p. 38) and the Coyote (p. 30). The largely nocturnal Eastern Cottontail can be seen at dawn or dusk and on darker days.

As with other rabbits and hares, this rabbit's most common track is a triangular grouping of four prints, with the larger hind prints (which can appear pointed) falling in front of the fore prints (which may overlap). The hairiness of the toes blurs the pad detail in the prints. If you follow this rabbit's tracks, the rabbit might startle you if it flies out from its 'form,' a depression in the ground in which it rests.

Similar Species: The White-tailed Jackrabbit (*Lepus townsendii*) leaves much larger print clusters and has longer strides, as does the European Hare (p. 70). The Snowshoe Hare (p. 66) also has larger prints, especially the hind ones. Tree squirrel (pp. 84–91) tracks show a similar pattern, but their fore prints are more consistently side by side.

European Hare

hind

fore

hopping

Fore Print
Length: 2–4 in (5–10 cm)
Width: 1.3 in (3.3 cm)

Hind Print
Length: 5 in (13 cm)
Length with heel: 9 in (23 cm)
Width: 2.5 in (6.5 cm)

Straddle
8 in (20 cm)

Stride
Hopping: 1.5–3 ft (45–90 cm)
Bounding: to 12 ft (3.7 m)

Size
Length: 25–30 in (65–75 cm)

Weight
7–12 lb (3.2–5.5 kg)

EUROPEAN HARE
Lepus europaeus

This large, sturdy colonizer of open fields was introduced to North America over one hundred years ago. It can now be found in the northern parts of these two states. Unlike the much smaller Snowshoe Hare (p. 66), which turns white in winter, the European Hare remains brown—and it has a black spot on its tail.

The usual track pattern of this hare, typical for rabbits and hares, consists of elongated triangular clusters of prints. The forefeet register first (usually in a slightly diagonal line), leaving small, roundish prints that show four toes. The two much larger hind prints then fall (usually side by side) ahead of the fore prints. The hind prints are much longer when the whole heel registers, as when the hare stops for a moment.

Similar Species: The smaller Eastern Cottontail (p. 68) makes smaller tracks, with a narrower straddle and smaller stride. The hind toes of the Snowshoe Hare splay out more in the snow.

Porcupine

fore

hind

Fore Print
Length: 2.3–3.3 in (5.8–8.5 cm)
Width: 1.3–1.9 in (3.3–4.8 cm)

Hind Print
Length: 2.8–4 in (7–10 cm)
Width: 1.5–2 in (3.8–5 cm)

Straddle
5.5–9 in (14–23 cm)

Stride
Walking: 5–10 in (13–25 cm)

Size
Length with tail: 25–40 in (65–100 cm)

Weight
10–28 lb (4.5–13 kg)

walking

PORCUPINE
Erethizon dorsatum

This notorious rodent rarely runs—its many long quills are a formidable defense. Common in the northern parts of Minnesota and Wisconsin, the Porcupine prefers forests, but it can also be seen in more open areas.

The Porcupine's preferred pigeon-toed, waddling gait leaves an alternating track pattern, with the hind print registered on or slightly in front of the shorter fore print. Look for long claw marks on all prints. The fore print shows four toes, and the hind print shows five. On clear prints, the unusual pebbly surface of the solid heel pads may show, but a Porcupine's tracks are often scratch-marked by its heavy, spiny tail. In deeper snow this squat animal drags its feet, and it may leave a trough with its body. A Porcupine's trail might lead you to a tree, where this animal spends much of its time feeding; if so, look for chewed bark or nipped twigs on the ground.

Similar Species: The Badger (p. 60) also has pigeon-toed prints, but it does not drag its tail or climb trees.

Beaver

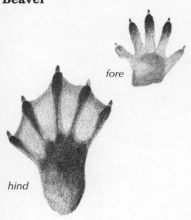

Fore Print
Length: 2.5–4 in (6.5–10 cm)
Width: 2–3.5 in (5–9 cm)
Hind Print
Length: 5–7 in (13–18 cm)
Width: 3.3–5.3 in (8.5–13 cm)
Straddle
6–11 in (15–28 cm)
Stride
Walking: 3–6.5 in (7.5–17 cm)
Size
Length with tail: 3–4 ft (90–120 cm)
Weight
28–75 lb (13–34 kg)

walking

BEAVER
Castor canadensis

Few animals leave as many signs of their presence as the Beaver, the largest North American rodent and a common sight around water. Look for the conspicuous dams and lodges—capable of changing the local landscape—and the stumps of felled trees. Check trunks gnawed clean of bark for marks of the Beaver's huge incisors. A scent mound marked with castoreum, a strong-smelling yellowish fluid that the Beaver produces, also indicates recent activity.

The Beaver's thick, scaly tail may mar its tracks, as can the branches that it drags about for construction and food. Check the large hind prints for signs of webbing and broad toenails—the nail of the second inner toe usually does not show. Rarely do all five toes on each foot register. Irregular foot placement in the alternating walking gait may produce a direct register or a double register. Repeated path use results in well-worn trails.

Similar Species: The Beaver's many signs, including its large hind prints, minimize confusion. Muskrat (p. 76) prints are smaller.

Muskrat

fore

hind

walking

Fore Print
Length: 1.1–1.5 in (2.8–3.8 cm)
Width: 1.1–1.5 in (2.8–3.8 cm)

Hind Print
Length: 1.6–3.2 in (4–8 cm)
Width: 1.5–2.1 in (3.8–5.3 cm)

Straddle
3–5 in (7.5–13 cm)

Stride
Walking: 3–5 in (7.5–13 cm)
Running: to 1 ft (30 cm)

Size
Length with tail: 16–25 in (40–65 cm)

Weight
2–4 lb (0.9–1.8 kg)

MUSKRAT
Ondatra zibethicus

Like the Beaver (p. 74), this rodent is found throughout the region, wherever there is water. Beavers are very tolerant of Muskrats and even allow them to live in parts of their lodges. Active year-round, the Muskrat leaves plenty of signs. It digs extensive networks of burrows, often undermining riverbanks, so do not be surprised if you suddenly fall into a hidden hole! Also look for small lodges in the water and beds of vegetation on which the Muskrat rests, suns and feeds in summer.

The small inner toe of the five on each forefoot rarely registers. The hind print shows five well-formed toes that may have a 'shelf' around them, created by stiff hairs that aid in swimming. The common alternating walking pattern shows print pairs that alternate from side to side, with the hind print just behind or slightly overlapping the fore print. In snow, a Muskrat's feet drag, and its tail leaves a sweeping dragline.

Similar Species: Few animals share this water-loving rodent's habits. Beavers make larger tracks and leave many other signs.

Woodchuck

fore

hind

Fore and Hind Prints
Length: 1.8–2.8 in (4.5–7 cm)
Width: 1–2 in (2.5–5 cm)

Straddle
3.3–6 in (8.5–15 cm)

Stride
Walking: 2–6 in (5–15 cm)
Bounding: 6–14 in (15–35 cm)

Size (male>female)
Length with tail:
  20–25 in (50–65 cm)

Weight
5.5–12 lb (2.5–5.5 kg)

walking   *bounding*

WOODCHUCK
(Whistle Pig, Groundhog, Marmot)
Marmota monax

This robust member of the squirrel family is a common sight in open woodlands and adjacent open areas throughout Minnesota and Wisconsin. Always on the watch for predators, but not too troubled by humans, the Woodchuck never wanders far from its burrow. This marmot hibernates during winter and emerges in early spring; look for tracks in late spring snowfalls and in the mud around the burrow entrances.

A Woodchuck's fore print shows four toes, three palm pads and two heel pads (not always evident). The hind print shows five toes, four palm pads and two poorly registering heel pads. The Woodchuck usually leaves an alternating walking pattern, with the hind print registered on the fore print. When a Woodchuck runs from danger, it makes groups of four prints, hind ahead of fore.

Similar Species: A small Raccoon's (p. 42) bounding track pattern will be similar, but it will show five-toed fore prints.

Thirteen-lined Ground Squirrel

fore

hind

Fore Print
Length: 1–1.3 in (2.5–3.3 cm)
Width: 0.5–1 in (1.3–2.5 cm)

Hind Print
Length: 1.1–1.5 in (2.8–3.8 cm)
Width: 0.8–1.3 in (2–3.3 cm)

Straddle
2.3–3.5 in (5.8–9 cm)

Stride
Bounding: 7–20 in (18–50 cm)

Size
Length with tail: 7–12 in (18–30 cm)

Weight
4–10 oz (110–280 g)

bounding

THIRTEEN-LINED GROUND SQUIRREL (Striped Gopher)
Spermophilus tridecemlineatus

With its many stripes and dots, this adorable ground squirrel of the shortgrass areas is an attractive sight. Its range extends throughout Minnesota and Wisconsin and into the Upper Midwest.

Look for this animal's tracks near its many burrow entrances, in mud or in late or early snowfalls. If there are no fresh tracks around the burrow, the animal may be dormant inside. The small fifth toe of the forefoot rarely registers in the print; the two heel pads sometimes show. The larger hind print shows five toes. Both fore and hind feet have long claws that often show in the prints. Ground squirrels are usually seen scurrying around, leaving a typical squirrel track pattern—the hind prints registering ahead of the fore prints, which are usually placed diagonally.

Similar Species: Franklin's Ground Squirrel (*S. columbianus*) is larger. Chipmunk (p. 82) tracks are smaller. Tree squirrel (pp. 84–91) tracks have a more square-shaped running group.

Eastern Chipmunk

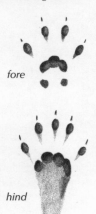

Fore Print
Length: 0.8–1 in (2–2.5 cm)
Width: 0.4–0.8 in (1–2 cm)
Hind Print
Length: 0.7–1.3 in (1.8–3.3 cm)
Width: 0.5–0.9 in (1.3–2.3 cm)
Straddle
2–3.2 in (5–8 cm)
Stride
Bounding: 7–15 in (18–38 cm)
Size
Length with tail: 7–10 in (18–25 cm)
Weight
2.5–5 oz (70–140 g)

bounding

EASTERN CHIPMUNK
Tamias striatus

This large chipmunk is found in a variety of habitats, from the dense forest floor to open areas near buildings. Look for this delightful character throughout Minnesota and Wisconsin. You are more likely to see or hear this rodent, which is highly active during summer, than to notice its tracks. This chipmunk is happiest on the ground, but it will gladly climb sturdy oak trees to harvest juicy, ripe acorns. Eastern Chipmunks enter a deep sleep in winter, waking up from time to time to have a meal.

Chipmunks are so light that their tracks rarely show fine details. The forefeet each have four toes and the hind feet five. Chipmunks run on their toes, so the two heel pads of the forefeet seldom register; the hind feet have no heel pads. Their erratic track patterns, like those of many of their cousins, show the hind feet registered in front of the forefeet. A chipmunk trail often leads to extensive burrows.

Similar Species: The Least Chipmunk (*Tamias minimus*) is smaller and less widespread. Tree squirrels (pp. 84–91) usually have larger prints and a wider straddle and are more likely to have made midwinter tracks. Mouse (pp. 98–101) tracks are smaller.

Eastern Gray Squirrel

fore

hind

Fore Print
Length: 1–1.8 in (2.5–4.5 cm)
Width: 1 in (2.5 cm)

Hind Print
Length: 2.3–3 in (5.8–7.5 cm)
Width: 1.1–1.5 in (2.8–3.8 cm)

Straddle
3.8–6 in (9.5–15 cm)

Stride
Bounding: 0.7–3 ft (22–90 cm)

Size
Length with tail: 17–20 in (43–50 cm)

Weight
14–25 oz (400–710 g)

bounding

EASTERN GRAY SQUIRREL
Sciurus carolinensis

This large and familiar squirrel can be a common sight in deciduous and mixed forests throughout most of this region, even in urban areas. Active all year, the Eastern Gray Squirrel can leave a wealth of evidence, especially in winter as it scurries about digging up nuts that it buried during the previous fall.

The Eastern Gray Squirrel leaves a typical squirrel track when it runs or bounds. The hind prints fall slightly ahead of the fore prints. A clear fore print shows four toes with sharp claws, four fused palm pads and two heel pads. The hind print shows five toes and four palm pads; if the full heel-length registers, it also shows two small heel pads.

Similar Species: Fox Squirrel (p. 86) prints are as big or larger. Red Squirrel (p. 88) prints are smaller. Eastern Chipmunks (p. 82) and flying squirrels (p. 90) make smaller tracks in a similar pattern and with narrower straddles. A rabbit or hare (pp. 66–71) makes a longer track pattern, and its forefeet rarely register side by side when it runs.

Fox Squirrel

fore

hind

Fore Print
Length: 1–1.9 in (2.5–4.5 cm)
Width: 1–1.7 in (2.5–4.3 cm)

Hind Print
Length: 2–3.3 in (5–7.5 cm)
Width: 1.5–1.9 in (3.8–4.8 cm)

Straddle
4–6 in (10–15 cm)

Stride
Bounding: 0.7–3 ft (22–90 cm)

Size
Length with tail: 18–28 in (45–70 cm)

Weight
1–2.4 lb (0.5–1.1 kg)

bounding

FOX SQUIRREL
Sciurus niger

This squirrel is much like the Eastern Gray Squirrel (p. 84), only larger and with a yellowish underside. It can be a common sight in deciduous forests with plenty of nut-bearing trees and in open areas or woodlands throughout Minnesota and Wisconsin. Piles of nutshells at tree bases indicate its favorite feeding sites. Active year-round, this squirrel spends a lot of time foraging on the ground, often collecting nuts that it buried singly during the previous fall.

The Fox Squirrel makes a typical squirrel track when it runs or bounds, with the hind prints appearing slightly in front of the fore prints, the prints in each pair roughly side by side. A clear fore print shows four toes with evident claws, four fused palm pads and two heel pads. The hind print shows five toes, four palm pads and sometimes a heel.

Similar Species: The Eastern Gray Squirrel generally leaves smaller prints. Chipmunk (p. 82) and flying squirrel (p. 90) tracks, in a similar pattern, are smaller with narrower straddles. Red Squirrel (p. 88) prints are much smaller. A rabbit or hare (pp. 66–71) makes a longer print pattern, and its fore prints rarely register side by side when it runs.

Red Squirrel

Fore Print
Length: 0.8–1.5 in (2–3.8 cm)
Width: 0.5–1 in (1.3–2.5 cm)
Hind Print
Length: 1.5–2.3 in (3.8–5.8 cm)
Width: 0.8–1.3 in (2–3.3 cm)
Straddle
3–4.5 in (7.5–11 cm)
Stride
Bounding: 8–30 in (20–75 cm)
Size
Length with tail:
    9–15 in (23–38 cm)
Weight
2–9 oz (55–260 g)

bounding

*bounding
(in deep snow)*

RED SQUIRREL
(**Pine Squirrel, Chickaree**)
Tamiasciurus hudsonicus

When you enter a
Red Squirrel's territory,
the inhabitant greets you
with a loud, chattering
call. Another obvious sign of
this forest dweller, which is found
throughout most of these two states,
is its large middens—piles of cone scales and cores
left beneath trees—that indicate favorite feeding sites.

Active year-round in their small territories, Red Squirrels leave an abundance of trails that lead from tree to tree or down a burrow. These energetic animals mostly bound, leaving groups of four prints; the hind prints appear in front of the fore prints, which tend to be side by side (but not always). Four toes show on each fore print, and five show on each hind print. The heels often do not register when squirrels move quickly. In deeper snow the prints merge to form pairs of diamond-shaped tracks.

Similar Species: The larger Eastern Gray Squirrel (p. 84) is also found throughout the region. Chipmunks (p. 82) and flying squirrels (p. 90) make tracks in a similar pattern, but have a narrower straddle and smaller prints.

Northern Flying Squirrel

fore

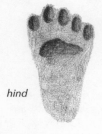
hind

Fore Print
Length: 0.5–0.8 in (1.3–2 cm)
Width: 0.5 in (1.3 cm)

Hind Print
Length: 1.3–1.8 in (3.3–4.5 cm)
Width: 0.8 in (2 cm)

Straddle
3–3.8 in (7.5–9.5 cm)

Stride
Bounding: 11–30 in (28–75 cm)

Size
Length with tail: 9–12 in (23–30 cm)

Weight
4–6.5 oz (110–180 g)

sitzmark into bounding

NORTHERN FLYING SQUIRREL
Glaucomys sabrinus

This acrobat can be found in coniferous forests throughout most of the region. It prefers widely spaced forests, where it can glide from tree to tree by night, using the membranous flaps between its legs. Northern Flying Squirrels will den up together in a tree cavity for warmth.

Because of its gliding, this squirrel does not leave as many tracks as other squirrels do. Evidence is scarce in summer, but in winter you may come across a sitzmark (the distinctive pattern that it made where it landed in the snow) and a short bounding trail, which it made as it rushed off to the nearest tree or to do some quick foraging. The bounding track pattern is typical of squirrels and other rodents, but with the hind feet registering only slightly in front of the forefeet—though often all four feet register in a row.

Similar Species: The Red Squirrel (p. 88) usually makes larger prints and rarely leaves a sitzmark, but unclear tracks in deep snow can be identical. Chipmunks (p. 82) make smaller prints with a narrower straddle.

Norway Rat

fore

hind

Fore Print
Length: 0.7–0.8 in (1.8–2 cm)
Width: 0.5–0.7 in (1.3–1.8 cm)

Hind Print
Length: 1–1.3 in (2.5–3.3 cm)
Width: 0.8–1 in (2–2.5 cm)

Straddle
2–3 in (5–7.5 cm)

Stride
Walking: 1.5–3.5 in (3.8–9 cm)
Bounding: 9–20 in (23–50 cm)

Size
Length with tail: 13–19 in (33–48 cm)

Weight
7–18 oz (200–510 g)

walking

NORWAY RAT
(Brown Rat)
Rattus norvegicus

Active both day and night, this despised rat is widespread almost anywhere that humans have decided to build their homes. Not entirely dependent on people, it may live in the wild as well.

The fore print shows four toes, and the hind print shows five. When it bounds, this colonial rat leaves four-print groups, with the hind prints in front of the diagonally placed fore prints. Sometimes one of the hind feet direct registers on a fore print, creating a three-print group. This rat more commonly leaves an alternating walking pattern with the larger hind prints close to or overlapping the fore prints; the hind heel does not show. The tail often leaves a dragline in snow. Rats live in groups, so you may find many trails together, often leading to their 5-inch (2-cm) wide burrows.

Similar Species: No other rat-like rodent of this size lives in this region. Mouse (pp. 98–101) prints are much smaller. Red Squirrel (p. 88) tracks show distinctive squirrel traits. Chipmunk (p. 82) tracks are smaller and the gaits differ.

Plains Pocket Gopher

fore

hind

Fore Print
Length: 1 in (2.5 cm)
Width: 0.6 in (1.5 cm)

Hind Print
Length: 0.8–1 in (2–2.5 cm)
Width: 0.5 in (1.3 cm)

Straddle
1.5–2 in (3.8–5 cm)

Stride
Walking: 1.3–2 in (3.3–5 cm)

Size (male> female)
Length with tail: 6–9 in (15–23 cm)

Weight
2.8–5 oz (80–140 g)

walking

PLAINS POCKET GOPHER
Geomys bursarius

    This seldom-seen rodent of the western and southern parts of this two-state region spends most of its time in burrows, only venturing out to move mud around and to find a mate. Because of its need to dig, the Plains Pocket Gopher prefers soft, moist soils, and it especially enjoys pastureland.

    By far the best signs by which to recognize pocket gopher activity are the muddy mounds and tunnel cores, which are most evident after the spring thaw. Each mound marks the entrance to a burrow, which is always blocked up with a plug. Search around a mound and you may find prints. Each foot has five toes. Although the gopher's forefeet have long, well-developed claws for digging, its prints rarely show this much detail. The Plains Pocket Gopher typically leaves an alternating walking track in which the hind prints fall on or slightly behind the fore prints.

Similar Species: The Plains Pocket Gopher's tracks are associated with its distinctive burrows, leaving little room for confusion.

Meadow Vole

fore

hind

Fore Print
Length: 0.3–0.5 in (0.8–1.3 cm)
Width: 0.3–0.5 in (0.8–1.3 cm)

Hind Print
Length: 0.3–0.6 in (0.8–1.5 cm)
Width: 0.3–0.6 in (0.8–1.5 cm)

Straddle
1.3–2 in (3.3–5 cm)

Stride
Walking/Trotting:
 1.3–3 in (3.3–7.5 cm)
Bounding: 4–8 in (10–20 cm)

Size
Length with tail:
 5.5–8 in (14–20 cm)

Weight
0.5–2.5 oz (14–70 g)

walking

*bounding
(in snow)*

MEADOW VOLE (Field Mouse)
Microtus pennsylvanicus

With so many vole species in the region, positive track identification is next to impossible, but note that the Meadow Vole frequents lush, damp or wet habitats.

When clear (which is seldom), vole fore prints show four toes, and hind prints show five. A vole's walk and trot both leave a paired alternating track pattern with a hind print occasionally direct registered on a fore print. Voles usually opt for a faster bounding; the resulting print pairs show the hind prints registered on the fore prints. These voles lope quickly across open areas, creating a three-print track pattern. Voles stay under the snow in winter; when it melts, look for distinctive piles of cut grass from their ground nests. The bark at the bases of shrubs may show tiny teeth marks left by gnawing. In summer, well-used vole paths appear as little runways in the grass.

Similar Species: The Southern Red-backed Vole (*Cletherionomys gapperi*) is found in central and northern parts of this area. The Prairie Vole (*M. ochrogaster*) is found in the south. Deer Mouse (p. 98) bounding tracks show four-print groups.

Deer Mouse

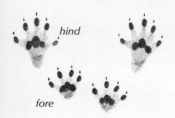
hind
fore
bounding group

bounding

bounding (in snow)

Fore Print
Length: 0.3–0.4 in (0.8–1 cm)
Width: 0.3–0.4 in (0.8–1 cm)
Hind Print
Length: 0.3–0.5 in (0.8–1.3 cm)
Width: 0.3–0.4 in (0.8–1 cm)
Straddle
1.4–1.8 in (3.6–4.5 cm)
Stride
Bounding: 5–12 in (13–30 cm)
Size
Length with tail:
    6–12 in (15–30 cm)
Weight
0.5–1.3 oz (14–35 g)

DEER MOUSE
Peromyscus maniculatus

The highly adaptable Deer Mouse–the most abundant mammal in this region–lives anywhere from arid valleys all the way up to alpine meadows. It is seldom seen because it is nocturnal. The Deer Mouse may enter buildings in winter, where it will stay active.

The fore prints each show four toes, three palm pads and two heel pads. The hind prints show five toes and three palm pads; the heel pads rarely register. It takes perfect, soft mud to get clear prints from such a tiny mammal. Bounding tracks, most noticeable in snow, show the hind prints falling in front of the fore prints. In soft snow the prints may merge to look like larger pairs of prints; tail drag will be evident. A mouse trail may lead up a tree or down into a burrow.

Similar Species: The tracks of many less common species of mice are identical. House Mouse (*Mus musculus*) tracks are very similar, but they are associated more with humans. Voles (p. 96) tend to trot and have a much shorter bounding track pattern. Jumping mouse (p. 100) prints may be similar in size, but show long, thin toes. Chipmunks (p. 82) have a wider straddle. Shrews (p. 102) have a narrower straddle.

Meadow Jumping Mouse

bounding group

Fore Print
Length: 0.3–0.5 in (0.8–1.3 cm)
Width: 0.3–0.5 in (0.8–1.3 cm)

Hind Print
Length: 0.5–1.3 in (1.3–3.3 cm)
Width: 0.5–0.7 in (1.3–1.8 cm)

Straddle
1.8–1.9 in (4.5–4.8 cm)

Stride
Bounding: 7–18 in (18–45 cm)
In alarm: 3–6 ft (90–180 cm)

Size
Length with tail: 7–9 in (18–23 cm)

Weight
0.6–1.3 oz (17–35 g)

bounding

MEADOW JUMPING MOUSE
Zapus hudsonius

Congratulations if you find and successfully identify the tracks of the Meadow Jumping Mouse! Though it lives throughout these two states, its preference for grassy meadows and its long, deep winter hibernation (about six months!) make locating tracks very difficult.

Jumping mouse tracks are distinctive if you do find them. The two smaller forefeet register between the long hind feet; the long heels do not always register and some prints show just the three long middle toes. The toes on the forefeet may splay so much that the side toes point backward. When they bound, these mice make short leaps. The tail may leave a dragline in soft mud or unseasonable snow. Clusters of cut grass stems about 5 inches (13 cm) long and lying in meadows are a more abundant sign of this rodent.

Similar Species: The Woodland Jumping Mouse (*Napaeozapus insignis*) is found in wet, wooded areas. Deer Mouse (p. 98) tracks may have the same straddle. Heel-less hind prints may be mistaken for a vole's (p. 96)—or a small bird's (p. 124) or an amphibian's (pp. 126–131).

Masked Shrew

hind

fore

bounding group

bounding

Fore Print
Length: 0.2 in (0.5 cm)
Width: 0.2 in (0.5 cm)
Hind Print
Length: 0.6 in (1.5 cm)
Width: 0.3 in (0.8 cm)
Straddle
0.8–1.3 in (2–3.3 cm)
Stride
Bounding: 1.2–3 in (3–7.5 cm)
Size
Length with tail: 2.3–4.5 in (7–11 cm)
Weight
0.1–0.3 oz (3–9 g)

MASKED SHREW
Sorex cinereus

Though several species of tiny, frenetic shrews are found in the region, the widespread and adaptable Masked Shrew is a likely candidate if you find tracks. This small shrew prefers moist fields, marshes, bogs or woodlands, but it can also be found in higher and drier grasslands. Its rapid activity makes it difficult to observe closely.

In its energetic and unending quest for food, a shrew usually leaves a four-print bounding pattern, but it may slow to an alternating walking gait. The individual prints in a group are often indistinct, but in mud or shallow, wet snow you can even count the five toes on each print. In deeper snow a shrew's tail often leaves a dragline. If a shrew tunnels under the snow, it may leave a snow ridge on the surface. A shrew's trail may disappear down a burrow.

Similar Species: The Pygmy Shrew (*S. hoyi*), more common in moist woods, and the Smoky Shrew (*S. fumeus*) leave similar tracks. The Water Shrew (*S. palustris*), often found by cold mountain streams, is larger; so is the widespread, abundant Northern Short-tailed Shrew (*Blarina brevicauda*). Mice (pp. 98–101) make four-toed fore prints.

Star-nosed Mole

a molehill of the Star-nosed Mole

some molehills and ridges of the Eastern Mole

Size
Length with tail: 6–8.5 in (15–22 cm)
Weight
1–2.6 oz (28–75 g)

STAR-NOSED MOLE
Condylura cristata

This peculiar character is found throughout most of Minnesota and Wisconsin. The Star-nosed Mole is easily identified by the strange tentacle-like protrusions on its nose, thought to help it find food in its dark and often subterranean world. It spends more time out of its burrow than other moles do.

Typical signs of this mole include the big piles of mud pushed out of its burrows. Because of its preference for swimming and wet areas, look for this mole's hills along the banks of streams and rivers and in raised areas around marshes and wet fields. Clear prints from its long-clawed feet are rarely found. The Star-nosed Mole remains under the snow during winter and swims under the ice, so winter tracks are seldom seen.

Similar Species: The Eastern Mole (*Scalopus aquaticus*), found in the extreme southeast of this two-state area, leaves the familiar distinctive molehills that are often connected by raised ridges created by digging tunnels near the surface (as illustrated at left).

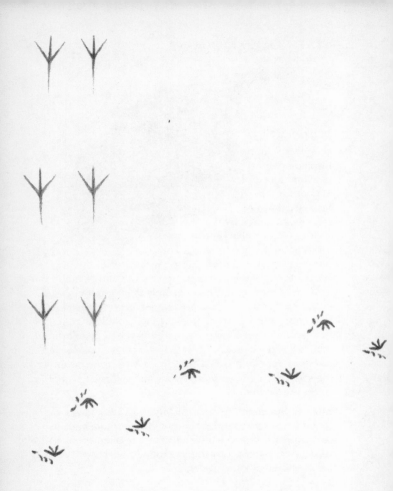

BIRDS, AMPHIBIANS & REPTILES

A guide to the animal tracks of Minnesota and Wisconsin is not complete without some consideration of the birds and amphibians found in the region.

Several bird species have been chosen to represent the main types common to this region. Remember, however, that individual bird species are not easily identified by track alone. Bird tracks can often be found in abundance in snow and are clearest in shallow, wet snow. The shores of streams and lakes are very reliable locations in which to find bird tracks—the mud there can hold a clear print for a long time. The sheer number of tracks made by shorebirds and waterfowl can be astonishing. Though some bird species prefer to perch in trees or soar across the sky, it can be entertaining to follow the tracks of birds that spend a lot of time on the ground. They can spin around in circles and lead you in all directions. The trail may suddenly end as the bird takes flight, or it might terminate in a pile of feathers, the bird having fallen victim to a hungry predator.

Many amphibians and turtles depend on moist environments, so look in the soft mud along the shores of lakes and ponds for their distinctive tracks. Though you may be able to distinguish frog tracks from toad tracks, because they generally move differently, it can be very difficult to identify the species. In drier environments, reptiles, which thrive in dryness, outnumber the amphibians, but they seldom leave good tracks.

Mallard

Print
Length: 2–2.5 in (5–6.5 cm)
Straddle
4 in (10 cm)
Stride
to 4 in (10 cm)
Size
23 in (58 cm)

MALLARD
Anas platyrhynchos

male

female

This dabbling duck—the male a familiar sight with its striking green head—is common in open areas near lakes and ponds. The Mallard's webbed feet leave prints that can often be seen in abundance along the muddy shores of just about any waterbody, including those in urban parks.

The webbed foot of the Mallard has three long toes that all point forward. Though the toes register well, the webbing between the toes does not always show in the print. The Mallard's inward-pointing feet give it a pigeon-toed appearance and perhaps account for its waddling gait, a characteristic for which ducks are known.

Similar Species: Other dabbling ducks and many gulls (p. 110) have similar prints. Exceptionally large prints are from geese or swans (*Cygnus* spp.).

Herring Gull

Print
Length: 3.5 in (9 cm)
Straddle
4–6 in (10–15 cm)
Stride
4.5 in (11 cm)
Size
Length: 23–25 in (58–65 cm)

HERRING GULL
Larus argentatus

The Herring Gull, with its long wings and webbed toes, is a strong long-distance flier as well as an excellent swimmer. Found throughout this region, it is concentrated in great numbers by the Great Lakes and other waterbodies.

Gulls leave slightly asymmetrical tracks that show three toes. They have claws that register outside the webbing, and the claw marks are usually attached to the footprint. Most gulls have quite a swagger to their gait, and they leave a trail with the tracks turned strongly inward.

Similar Species: Gull species cannot be reliably identified by track alone, but smaller species have conspicuously smaller tracks. Duck tracks (p. 108) are difficult to distinguish from gull tracks.

Great Blue Heron

Print
Length: to 6.5 in (17 cm)

Straddle
8 in (20 cm)

Stride
9 in (23 cm)

Size
4.2–4.5 ft (1.3–1.4 m)

GREAT BLUE HERON
Ardea herodias

The refined and graceful image of this large heron symbolizes the precious wetlands in which it patiently hunts for food. Usually still and statuesque as it waits for a meal to swim by, the Great Blue Heron will have cause to walk from time to time, perhaps to find a better hunting location. Look for its large, slender tracks along the banks or mudflats of waterbodies.

Not surprisingly, a bird that lives and hunts with such precision walks in a similar fashion, leaving straight tracks that fall in a nearly straight line. Look for the slender rear toe in the print.

Similar Species: Cranes (*Grus* spp.), which occupy similar habitats, have similar and possibly larger tracks, but a crane's rear toes are smaller and do not register.

Common Snipe

Print
Length: 1.5 in (3.8 cm)
Straddle
to 1.8 in (4.5 cm)
Stride
to 1.3 in (3.3 cm)
Size
11–12 in (28–30 cm)

COMMON SNIPE
Gallinago gallinago

This short-legged character is a resident of marshes and bogs, where its neat prints can often be seen in mud. Snipes are quite secretive when on the ground, so you may be surprised if one suddenly flushes out from beneath your feet. If there is a Common Snipe in the air, you may hear an eerie whistle as it dives from the sky.

The Common Snipe's neat prints show four toes, including a small rear toe that points inward. The bird's short legs and stocky body give it a very short stride.

Similar Species: Many shorebirds, including the Spotted Sandpiper (p. 116), leave similar tracks.

Spotted Sandpiper

Print
Length: 0.8–1.3 in (2–3.3 cm)

Straddle
to 1.5 in (3.8 cm)

Stride
Erratic

Size
7–8 in (18–20 cm)

SPOTTED SANDPIPER
Actitis macularia

The bobbing tail of the Spotted Sandpiper is a common sight on the shores of lakes, rivers and streams, but you will usually find just one of these territorial birds in any given location. Because of its excellent camouflage, likely the first that you will see of this bird will be when it flies away, its fluttering wings close to the surface of the water.

As a sandpiper teeters up and down on the shore, it leaves trails of three-toed prints. Its fourth toe is very small and faces off to one side at an angle. Sandpiper tracks can have an erratic stride.

Similar Species: All sandpipers and plovers, including the Killdeer (*Charadrius vociferus*), leave similar tracks, although there is much diversity in size. The Common Snipe (p. 114) makes larger tracks.

Ruffed Grouse

Print
Length: 2–3 in (5–7.5 cm)
Straddle
2–3 in (5–7.5 cm)
Stride
Walking: 3–6 in (7.5–15 cm)
Size
15–19 in (38–48 cm)

RUFFED GROUSE
Bonasa umbellus

This ground-dweller prefers the quiet seclusion of coniferous forests in winter, so that will be the best place to find its tracks. If you follow them quietly, you may be startled when the Ruffed Grouse bursts from cover underneath your feet. Its excellent camouflage usually affords it good protection.

The three thick front toes leave very clear impressions, but the short rear toe, which is angled off to one side, does not always show up so well. This bird's neat, straight trail appears to reflect its cautious approach to life on the forest floor.

Similar Species: Other grouse leave similar tracks, but their prints may be blurred and enlarged by the winter feathers that they grow on their feet. The Ring-necked Pheasant (*Phasianus colchicus*) makes similar but larger tracks.

Great Horned Owl

Strike
Width: to 3 ft (90 cm)
Size
22 in (55 cm)

GREAT HORNED OWL
Bubo virginianus

Often seen resting quietly in trees by day, this wide-ranging owl prefers to hunt at night. An accomplished hunter in snow, the owl strikes through the snow with its talons, leaving an untidy hole that may be surrounded by wing and tail-feather imprints. If it registers well, this 'strike' can be quite a sight. The feather imprints are made as the owl struggles to take off with possibly heavy prey. An ungraceful walker, it prefers to fly away from the scene.

You may stumble across a strike and guess that the owl's target could have been a vole (p. 96) scurrying around underneath the snow. Or you may be following the surface trail of an animal to find that it abruptly ends with this strike mark, where the animal has been seized.

Similar Species: If the prey left no approaching trail, the strike mark is likely an owl's, because owls hunt by sound. If there is a trail, the strike mark, usually with less-rounded and more distinct feather imprints, could be by a hawk or a Common Raven (*Corvus corax*), both of whom hunt by sight.

American Crow

Print
Length: 2.5–3 in (6.5–7.5 cm)
Straddle
1.5–3 in (3.8–7.5 cm)
Stride
Walking: 4 in (10 cm)
Size
16 in (40 cm)

AMERICAN CROW
Corvus brachyrhyncos

The black silhouette of the American Crow can be a common sight in a variety of habitats. A crow will frequently come down to the ground and contentedly strut around with a confidence that hints at its intelligence. Its loud *caw* can be heard from quite a distance; crows can be especially noisy when they are mobbing an owl or a hawk.

The American Crow typically leaves an alternating walking track pattern. Its prints show three sturdy toes pointing forward and one toe pointing backward. When a crow is in need of greater speed, perhaps for takeoff, it bounds along, leaving irregular pairs of diagonally placed prints with a longer stride between each pair.

Similar Species: Other corvids, such as jays, also spend a lot of time on the ground and make similar tracks. The much larger Common Raven (*C. corax*), which may occur in forested regions of these two states, leaves tracks to 4 inches (10 cm) long and with a stride of 6 inches (15 cm).

Dark-eyed Junco

Print
Length: to 1.5 in (3.8 cm)
Straddle
1–1.5 in (2.5–3.8 cm)
Stride
Hopping: 1.5–5 in (3.8–13 cm)
Size
5.5–6.5 in (14–17 cm)

DARK-EYED JUNCO
Junco hyemalis

This common small bird typifies the many small hopping birds found in the region. Each foot has three forward-pointing toes and one longer toe at the rear. The best prints are left in snow, although in deep snow the toe detail is lost; the footprints may show some dragging between the hops.

A good place to study this type of prints is near a birdfeeder. Watch the birds scurry around as they pick up fallen seeds, then have a look at the prints left behind. For example, Dark-eyed Juncos are attracted to seeds that chickadees (*Poecile* spp.) scatter as they forage for sunflower seeds in the birdfeeder. Also look for tracks under coniferous trees, where juncos feed on fallen seeds in winter.

Similar Species: Toe size may help with identification—larger birds make larger prints—as can the season. In powdery snow, junco tracks could be mistaken for mouse (pp. 98–101) tracks, so follow the trail to see if the tracks disappear down a hole or into thin air.

Frogs

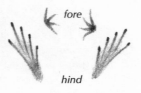

fore

hind

hopping

Straddle
to 3 in (7.5 cm)

FROGS

Wood Frog

The best place to look for frog tracks is along the muddy fringes of waterbodies.

The smallest frogs include the treefrogs. The Spring Peeper's (*Pseudoacris crucifer*) 1.5-inch (3.8-cm) length and its preference for thick undergrowth and shrubs near the water make its tracks a rare sight. The larger Gray Treefrogs (*Hyla chrysoscelis* and *H. versicolor*) spend most of their time in trees, coming down to breed and sing at night. The widespread Wood Frog (*Rana sylvatica*), often found in dry woodland areas seemingly far from water, is larger, at about 3.5 inches (9 cm) long. The Green Frog (*R. clamitans*) and the Pickerel Frog (*R. palustris*) are also widespread. The beautiful and widespread Northern Leopard Frog (*R. pipiens*), to 5 inches (13 cm) long, makes larger tracks. Unusually large tracks are surely from the robust Bullfrog (*R. catesbeiana*). Growing to 8 inches (20 cm) in length, it is North America's largest frog.

A frog's hopping action results in its two small forefeet registering in front of its long-toed hind prints. Frog tracks vary greatly in size, depending on species and age. Toads (p. 128) may also hop, but usually they walk.

Toads

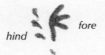
hind *fore*

Straddle
to 2.5 in (6.5 cm)

walking

TOADS

Woodhouse's Toad

There are fewer toad species than frog species in Minnesota and Wisconsin. The best place to look for toad tracks is, as with frogs, along the muddy fringes of waterbodies, but their tracks can also be found in drier areas, as unclear trails in dusty patches of soil.

The toad most likely to be encountered, and the most widespread, is the American Toad (*Bufo americanus*), which lives in many different moist habitats. Woodhouse's Toad (*B. woodhousei*) is found scattered throughout the region in temporary pools and ditches. Toads in this region can be up to 4.5 inches (11 cm) in length.

In general, toads walk and frogs (p. 126) hop, but toads are pretty capable hoppers, too, especially when being hassled by overly enthusiastic naturalists. Toads leave rather abstract prints as they walk. The heels of the hind feet do not register. On less-firm surfaces, the toes often leave draglines.

Salamanders & Newts

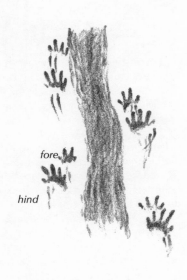

Straddle
to 3 in (7.5 cm)

walking

SALAMANDERS & NEWTS

Redback Salamander

There are a wealth of salamanders and newts in the moist and wet areas of Minnesota and Wisconsin. Among the more abundant and widespread of these long, slender, lizard-like amphibians is the Eastern Newt (*Notophthalmus viridescens*), which can grow to 5.5 inches (14 cm) in length. After a fresh rain, Eastern Newts emerging from ponds may leave small trails in the mud.

Other species of salamanders outnumber the newts in this region. The Redback Salamander (*Plethodon cinereus*), which can reach 5 inches (13 cm) in length, lives throughout the region in mixed forests and some coniferous forests. The king of the salamander world is undoubtedly the magnificent Tiger Salamander (*Ambystoma tigrinum*), which can grow up to 13 inches (33 cm) long and comes in such a diversity of colors and pattern that it defies a worthy description. This heavy salamander leaves highly visible tracks with a straddle of up to 4 inches (10 cm).

In general, a salamander or newt fore print shows four toes, and the larger hind print shows five. However, print detail is often blurred by the animal's dragging belly or by the swinging of its thick tail across the tracks.

Lizards

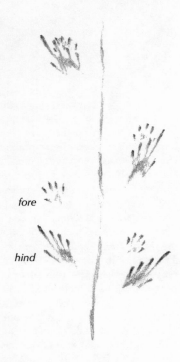

fore

hind

Straddle
to 3 in (7.5 cm)

walking

LIZARDS

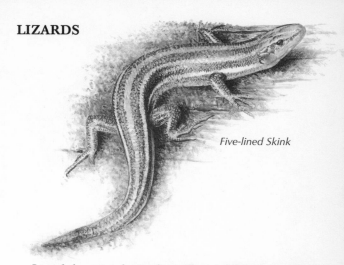

Five-lined Skink

Lizards leave tracks similar to those of the salamanders, but their toes are longer and more slender. If you find lizard tracks, the most likely candidate is the Five-lined Skink (*Eumeces fasciatus*), which can grow up to 8 inches (20 cm) in length. It favors moist woodlands and is widespread throughout these two states. The other skink that you might encounter is the Prairie Skink (*E. septentrionalis*).

Other lizards are uncommon in the region, but you may find tracks of the Racerunner (*Cnemidophorus sexlineatus*). This lizard, which can reach 11 inches (28 cm) in length, favors dry grasslands or well-drained woodlands.

These reptiles all move very quickly when the need arises, their feet barely touching the ground as they dart for cover. Consequently, their tracks can be hard to make out.

Turtles

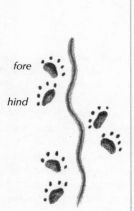
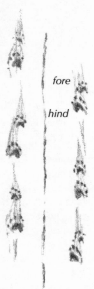

Straddle
4–10 in (10–25 cm)

Snapping Turtle walking

typical turtle walking

TURTLES

Eastern Box Turtle

Turtles, those ancient inhabitants of the water world, will happily slip into the murky depths of the water to avoid detection. They do, however, come out from time to time to feed or to bask in the sunshine. Look for their distinctive tracks alongside ponds, rivers and moist areas. Note that some turtles, such as the huge Snapping Turtle (*Chelydra serpentina*), prefer to stay in the water and rarely come out.

One of the most widespread and prettiest turtles, often seen basking, is the Painted Turtle (*Chrysemys picta*); it can reach almost 10 inches (25 cm) in length. Slightly smaller is the Eastern Box Turtle (*Terrapene carolina*), which likes moist forested areas. The Map Turtle (*Graptemys geographica*), which reaches only 6 inches (15 cm) in length, prefers slow-moving rivers.

With its large shell and short legs, a turtle leaves a track that is wide relative to the length of its stride—its straddle is about half its body length. Although longer-legged turtles can raise their shells off the ground, short-legged species may let their shells drag, as shown in their tracks. The tail may leave a straight dragline in the mud. On firmer surfaces, look for distinct claw marks.

Snakes

typical snake

SNAKES

Common Garter Snake

There are many snakes to be found throughout Minnesota and Wisconsin, with much greater diversity in the warmer south. Because snakes are all long and slender, their tracks appear so similar that identification among the species is next to impossible. In fact, because a snake lacks feet and leaves a track that is just a gentle meander, it is very challenging even to establish in which direction a snake was traveling.

The most widespread and frequently encountered snake is the Common Garter Snake (*Thamnophis sirtalis*). Found throughout the region, often close to wet or moist areas, this harmless snake can reach 4.3 feet (1.3 m) in length.

Another widespread species is the Red-bellied Snake (*Storeria occipitomaculata*), which prefers hilly woodlands and reaches a length of up to 16 inches (40 cm). Only slightly larger, the Smooth Green Snake (*Opheodrys vernalis*) inhabits grassy meadows and fields along forest edges. The rattlesnake most likely to be encountered is the Massasauga (*Sistrurus catenatus*); it can reach a length of 3.3 feet (1 m) and frequents a variety of habitats from marshlands to dry woodlands. In sandy, open areas, a snake track might be from the Eastern Hognose Snake (*Heterodon platirhinos*).

TRACK PATTERNS & PRINTS

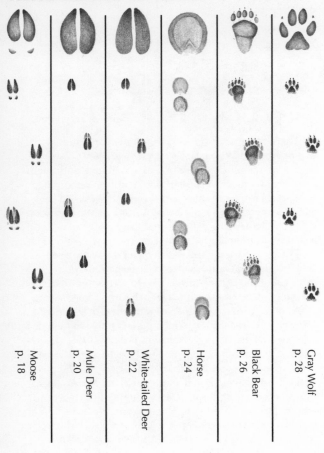

Moose p. 18

Mule Deer p. 20

White-tailed Deer p. 22

Horse p. 24

Black Bear p. 26

Gray Wolf p. 28

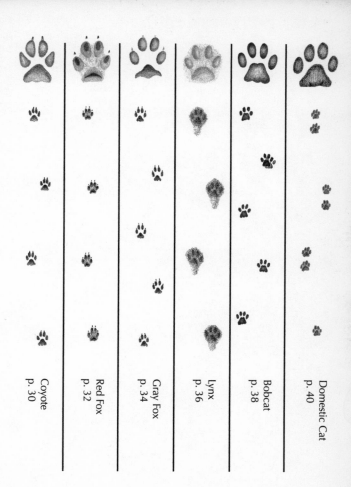

| Coyote p. 30 | Red Fox p. 32 | Gray Fox p. 34 | Lynx p. 36 | Bobcat p. 38 | Domestic Cat p. 40 |

TRACK PATTERNS & PRINTS

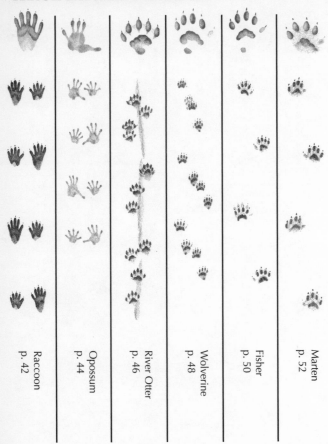

Raccoon p. 42

Opossum p. 44

River Otter p. 46

Wolverine p. 48

Fisher p. 50

Marten p. 52

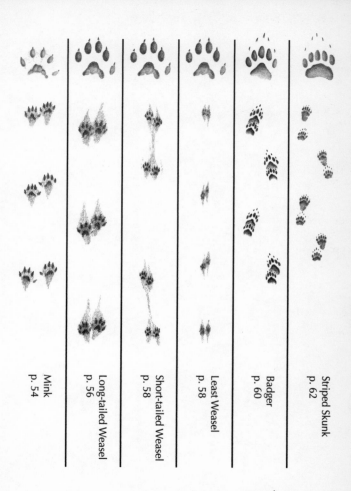

TRACK PATTERNS & PRINTS

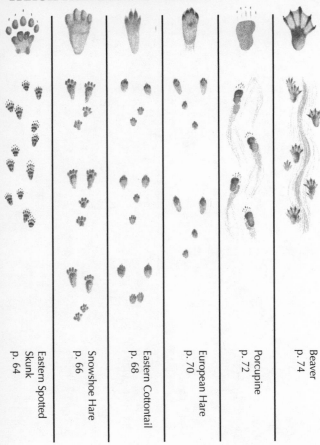

Eastern Spotted Skunk p. 64

Snowshoe Hare p. 66

Eastern Cottontail p. 68

European Hare p. 70

Porcupine p. 72

Beaver p. 74

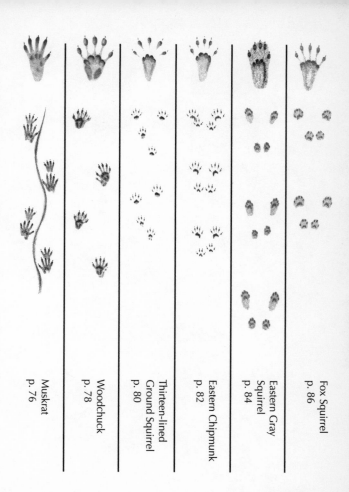

TRACK PATTERNS & PRINTS

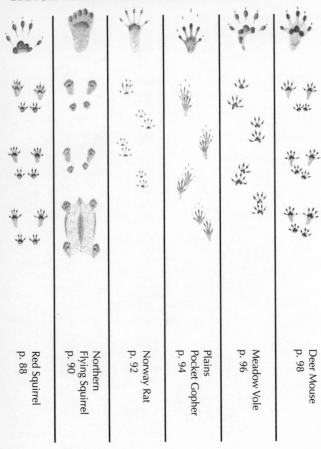

Red Squirrel p. 88

Northern Flying Squirrel p. 90

Norway Rat p. 92

Plains Pocket Gopher p. 94

Meadow Vole p. 96

Deer Mouse p. 98

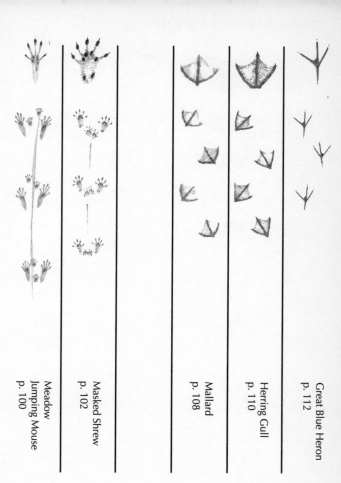

TRACK PATTERNS & PRINTS

Common Snipe
p. 114

Spotted Sandpiper
p. 116

Ruffed Grouse
p. 118

American Crow
p. 122

Dark-eyed Junco
p. 124

| Snakes p. 136 | Turtles p. 134 | Lizards p. 132 | Salamanders & Newts p. 130 | Toads p. 128 | Frogs p. 126 |

HOOFED PRINTS

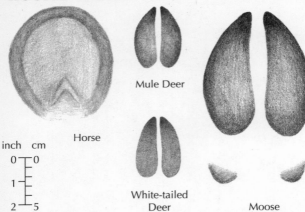

HIND PRINTS

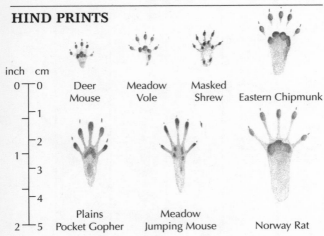

HIND PRINTS

Thirteen-lined Ground Squirrel

Northern Flying Squirrel

Red Squirrel

Fox Squirrel

Eastern Gray Squirrel

Woodchuck

Opossum

Raccoon

Porcupine

Muskrat

Eastern Cottontail

Snowshoe Hare

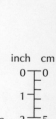
European Hare

HIND PRINTS

inch cm
0 — 0
2
4 — 10

Beaver

Black Bear

FORE PRINTS

Domestic Cat

Bobcat

Lynx

Gray Fox

Red Fox

inch cm
0 — 0
1
2 — 5

Coyote

Gray Wolf

FORE PRINTS

Least Weasel   Short-tailed Weasel   Long-tailed Weasel   Eastern Spotted Skunk   Striped Skunk

Mink   Marten   River Otter   Fisher

Badger   Wolverine

BIBLIOGRAPHY

Behler, J.L., and F.W. King. 1979. *Field Guide to North American Reptiles and Amphibians.* National Audubon Society. New York: Alfred A. Knopf.

Brown, R., J. Ferguson, M. Lawrence and D. Lees. 1987. *Tracks and Signs of the Birds of Britain and Europe: An Identification Guide.* London: Christopher Helm.

Burt, W.H. 1976. *A Field Guide to the Mammals.* Boston: Houghton Mifflin Company.

Farrand, J., Jr. 1995. *Familiar Animal Tracks of North America.* National Audubon Society Pocket Guide. New York: Alfred A.-Knopf.

Forrest, L.R. 1988. *Field Guide to Tracking Animals in Snow.* Harrisburg: Stackpole Books.

Halfpenny, J. 1986. *A Field Guide to Mammal Tracking in North America.* Boulder: Johnson Publishing Company.

Headstrom, R. 1971. *Identifying Animal Tracks.* Toronto: General Publishing Company.

Murie, O.J. 1974. *A Field Guide to Animal Tracks.* The Peterson Field Guide Series. Boston: Houghton Mifflin Company.

Rezendes, P. 1992. *Tracking and the Art of Seeing: How to Read Animal Tracks and Sign.* Vermont: Camden House Publishing.

Stall, C. 1989. *Animal Tracks of the Rocky Mountains.* Seattle: The Mountaineers.

Stokes, D., and L. Stokes. 1986. *A Guide to Animal Tracking and Behaviour.* Toronto: Little, Brown and Company.

Wassink, J.L. 1993. *Mammals of the Central Rockies.* Missoula: Mountain Press Publishing Company.

Whitaker, J.O., Jr. 1996. *National Audubon Society Field Guide to North American Mammals.* New York: Alfred A. Knopf.

INDEX

Actitis macularia, **116**
Alces alces, **18**
Ambystoma tigrinum, 131
Amphibians, 101, 107, **126–131**
Anas platyrhynchos, **108**
Ardea herodias, **112**

Badger, 31, **60**, 73
Bear
  Black, **26**
  Brown. *See* Grizzly B.
  Grizzly, 27
Beaver, **74**, 77
Birds, 101, 107, **108–125**
Blarina brevicauda, 103
Bobcat, 35, 37, **38**, 41, 51, 69
Bonasa umbellus, **118**
Bubo virginianus, **120**
Bufo
  americanus, 129
  woodhousei, **128**
Bullfrog, 127

Canids, **28–35**
Canis
  familiaris, 29, 31, 33, 37, 39, 41, 49

Canis (cont.)
  latrans, **30**
  lupus, **28**
Castor canadensis, **74**
Cat
  Black. *See* Fisher
  Domestic, 35, 39, **40**
  House. *See* Domestic C.
Charadrius vociferus, 117
Chelydra serpentina, 135
Chickadees, 125
Chickaree. *See* Squirrel, Red
Chipmunk, 81, 87, 89, 91, 99
  Eastern, **82**, 85
  Least, 83
Chrysemys picta, 135
Cletherionomys gapperi, 97
Cnemidophorus sexlineatus, 133
Condylura cristata, **104**
Corvids, **122–123**
Corvus
  brachyrhyncos, **122**
  corax, 121, 123
Cottontail, Eastern, 67, **68**, 71
Coyote, 29, **30**, 33, 35, 37, 39, 67, 69
Cranes, 113

153

Crow, American, **122**
Cygnus spp., 109

Deer, 19
　Mule, **20**, 23
　White-tailed, 21, **22**
Didelphis virginiana, **44**
Dog, Domestic, 29, 31, 33, 37, 39, 41, 49
Duck, 111
　Dabbling, 109
　Mallard, **108**

Equus caballus, **24**
Erethizon dorsatum, **72**
Ermine.
　See Weasel, Short-tailed
Eumeces
　fasciatus, **132**
　septentrionalis, 133

Felis catus, **40**
Fisher, 37, 43, 47, **50**, 53
Flying Squirrel, 85, 87, 89
　Northern, **90**
Fox, 37, 39, 41
　Gray, 31, 33, **34**
　Red, 31, **32**, 35
Frog, 129
　Bullfrog, 127
　Gray Treefrogs, 127

Frog (*cont.*)
　Green, 127
　Northern Leopard, 127
　Pickerel, 127
　Spring Peeper, 127
　Wood, **126**

Gallinago gallinago, **114**
Geese, 109
Geomys bursarius, **94**
Glaucomys sabrinus, **90**
Gopher, Plains Pocket, **94**
Gopher, Striped. *See* Ground Squirrel, Thirteen-lined
Graptemys geographica, 135
Ground Squirrel
　Franklin's, 81
　Thirteen-lined, **80**
Groundhog. *See* Woodchuck
Grouse, Ruffed, **118**
Grus spp., 113
Gull, 109
　Herring, **110**
Gulo gulo, **48**

Hare, 55, 57, 85, 87
　European, 69, **70**
　Snowshoe, 37, **66**, 69, 71
　Varying. *See* Snowshoe H.
Hawks, 121
Heron, Great Blue, **112**

Heterodon platirhinos, 137
Horse, **24**
Hyla
  chrysoscelis, 127
  versicolor, 127

Jackrabbit, White-tailed, 67, 69
Jays, 123
Jumping Mouse, 99
  Meadow, **100**
  Woodland, 101
Junco hyemalis, **124**
Junco, Dark-eyed, **124**

Killdeer, 117

Larus argentatus, **110**
Lepus
  americanus, **66**
  europaeus, 69, **70**
  townsendii, 67, 69
Lizard, Racerunner, 133
Lontra canadensis, **46**
Lynx, **36**, 67
Lynx
  canadensis, **36**
  rufus, **38**

Mallard, **108**
Marmot. *See* Woodchuck

Marmota monax, **78**
Marten, 47, 51, **52**, 55
Martes
  americana, **52**
  pennanti, **50**
Massasauga, 137
Mephitis mephitis, **62**
Microtus
  ochrogaster, 97
  pennsylvanicus, **96**
Mink, 47, 53, **54**, 57
Mole
  Eastern, 105
  Star-nosed, **104**
Moose, **18**, 21, 23
Mouse, 83, 93, 103, 125
  Deer, 97, **98**, 101
  Field. *See* Vole, Meadow
  House, 99
  Jumping, 99
  Meadow Jumping, **100**
  Woodland Jumping, 101
Mule, 25
Mus musculus, 99
Muskrat, 75, **76**
Mustela
  erminea, **58**
  frenata, **56**
  nivalis, **58**
  vison, **54**
Mustelids, 37, 39, **46–61**, 63

Napaeozapus insignis, 101
Newt, Eastern, 131
Notophthalmus viridescens, 131

Odocoileus
  hemionus, **20**
  virginianus, **22**
Ondatra zibethicus, **76**
Opheodrys vernalis, 137
Opossum, 43, **44**
Otter, River, 43, **46**, 51, 53, 55
Owl, Great Horned, **120**

Peromyscus maniculatus, **98**
Phasianus colchicus, 119
Pheasant, Ring-necked, 119
Plethodon cinereus, **130**
Plovers, 117
Pocket Gopher, Plains, **94**
Poecile spp., 125
Porcupine, 51, 61, **72**
Procyon lotor, **42**
Pseudoacris crucifer, 127

Rabbits, 55, 57, 67, 69, 71, 85, 87
Raccoon, **42**, 45, 79
Racerunner, 133

Rana
  catesbeiana, 127
  clamitans, 127
  pipiens, 127
  sylvatica, **126**
Rat
  Brown. *See* Norway R.
  Norway, **92**
Rattlesnake, Massasauga, 137
Rattus norvegicus, **92**
Raven, Common, 121, 123
Reptiles, 107, **132–137**

Sable, American. *See* Marten
Salamander
  Redback, **130**
  Tiger, 131
Sandpiper, Spotted, 115, **116**
Scalopus aquaticus, 105
Sciurus
  carolinensis, **84**
  niger, **86**
Shorebirds, 115
Shrew, 99
  Masked, **102**
  Northern Short-tailed, 103
  Pygmy, 103
  Smoky, 103
  Water, 103
Sistrurus catenatus, 137

Skink
  Five-lined, **132**
  Prairie, 133
Skunk
  Eastern Spotted, 63, **64**
  Striped, **62**, 65
Snake
  Common Garter, **136**
  Eastern Hognose, 137
  Massasauga, 137
  Red-bellied, 137
  Smooth Green, 137
Snipe, Common, **114**, 117
Sorex
  cinereus, **102**
  fumeus, 103
  hoyi, 103
  palustris, 103
Spermophilus
  columbianus, 81
  tridecemlineatus, **80**
Spilogale putorius, **64**
Spring Peeper, 127
Squirrel
  Eastern Gray, **84**, 87, 89
  Flying, 85, 87, 89
  Fox, 85, **86**
  Franklin's Ground, 81
  Northern Flying, **90**
  Pine. *See* Red S.

Squirrel (*cont.*)
  Red, 85, 87, **88**, 91, 93
  Thirteen-lined Ground, **80**
  Tree, 69, 81, 83
Stoat.
  See Weasel, Short-tailed
Storeria occipitomaculata, 137
Swans, 109
Sylvilagus floridanus, **68**

Tamias
  minimus, 83
  striatus, **82**
Tamiasciurus hudsonicus, **88**
Taxidea taxus, **60**
Terrapene carolina, **134**
Thamnophis sirtalis, **136**
Toad, 127
  American, 129
  Woodhouse's, **128**
Treefrogs, Gray, 127
Turtle
  Eastern Box, **134**
  Map, 135
  Painted, 135
  Snapping, 135

Urocyon cinereoargenteus, **34**
Ursus
  americanus, **26**
  arctos, 27

Vole, 99, 101, 121
 Meadow, **96**
 Prairie, 97
 Southern Red-backed, 97
Vulpes vulpes, **32**

Weasel, 55
 Least, 57, **58**
 Long-tailed, **56**, 59
 Short-tailed, 57, **58**
Whistle Pig. *See* Woodchuck
Wildcat. *See* Bobcat

Wolf
 Brush. *See* Coyote
 Gray, **28**, 49
 Prairie. *See* Coyote
 Timber. *See* Gray W.
Wolverine, 29, **48**
Woodchuck, **78**

Zapus hudsonius, **100**

Gray Wolf

ABOUT THE AUTHORS

Ian Sheldon, an accomplished artist, naturalist and educator, has lived in South Africa, Singapore, Britain and Canada. Caught collecting caterpillars at the age of three, he has been exposed to the beauty and diversity of nature ever since. He was educated at Cambridge University and the University of Alberta. When he is not in the tropics working on conservation projects or immersing himself in our beautiful wilderness, he is sharing his love for nature. Ian enjoys communicating this passion through the visual arts and the written word.

Tamara Eder, equipped from the age of six with a canoe, a dip net, and a note pad, grew up with a fascination for nature and the diversity of life. She has a degree in environmental conservation sciences and has photographed and written about the biodiversity in Bermuda, the Galapagos Islands, the Amazon Basin, China, Tibet, Vietnam, Thailand and Malaysia.

Explore the World Outside Your Door!

Mushrooms of Northeast North America, Midwest to New England
By George Barron
ISBN 978-1-55105-201-4 • 336 Pages • Over 600 color photographs
$24.95

A full-color photographic field guide to more than 600 species of mushrooms and fungi of the northern United States, from the Midwest to New England. Spectacular photographs and excellent species information combine to make this a must-have reference.

Weeds of the Northern U.S. and Canada
By Richard Dickinson & France Royer
ISBN 978-1-55105-221-2 • 472 pages • 800 color photographs
$21.95

This impressive, richly illustrated field guide identifies more than 150 noxious weeds. Color photographs show the weeds at the five critical stages and help to distinguish another 100 related species. Information on weed legislation by state is also listed. A first of its kind, this book will be an extraordinary resource for a multitude of users, whether farmer, landscaper, weed specialist or gardener.

Birds of Chicago
By Chris C. Fisher & David B. Johnson
ISBN 978-1-55105-182-6 • 160 Pages • Color illustrations
$9.95

This easy-to-use and beautifully illustrated guidebook will help you identify the feathered strangers nibbling at the feeder in your backyard or singing from a nearby tree. It's packed with notes on 126 common and interesting bird species in and around Chicago, including NE Illinois and NW Indiana.

US Orders	**Canadian Orders**
1-800-518-3541 Phone	1-800-661-9017 Phone
1-800-548-1169 Fax	1-800-424-7173 Fax

E-mail: heleni@wolfenet.com